**International
Business
and Global
Technology**

International Business and Global Technology

J. Davidson Frame
The George Washington University

LexingtonBooks
D.C. Heath and Company
Lexington, Massachusetts
Toronto

Library of Congress Cataloging in Publication Data

Frame, J. Davidson.
 International business and global technology.

 Includes index.
 1. Technology transfer. 2. International business enterprises. I. Title.
T174.3.F73 1982   338.9   82-48480
ISBN 0-669-06156-5

Published simultaneously in Canada

Printed in the United States of America

International Standard Book Number: 0-669-06156-5

Library of Congress Catalog Card Number: 82-48480

To my parents

Contents

Figures and Tables

Preface and Acknowledgments

This book discusses two of the great forces of our day. One is the irreversible growing interdependence of all countries, as reflected in the increasingly complex nexus of transnational commercial relations. The other is the inexorable advance of science and technology, which continuously changes the character of our world and lives. Of course, these two forces are not independent. Technological advances create a shrinking globe and make possible increasingly complex ties among countries. Conversely, the growth of the international marketplace stimulates technological advances via technology transfer and competition-induced innovation.

Despite the obvious link between these two forces, for the most part each lies in a distinct realm. The first force is principally the province of politicians, bureaucrats, bankers, and businessmen. The second lies in the province of scientists and engineers. As C.P. Snow pointed out several decades ago, our society lends itself easily to the existence of two cultures—the culture of scientists and engineers who speak their own language and in certain respects constitute a modern-day priestly caste, and the culture of everyone else. We see the existence of these two cultures when we look at international business activities today. On the one hand are international business people who are familiar with international finance, marketing, trade, and so forth. These people often have little substantive appreciation of the crucial role of technology in driving international business forward. On the other hand are scientists and engineers who are able to perform miracles with electronics, chemistry, and biology. Too often, the machinations of international business are as mysterious to them as scientific formulas are to laymen, and they are unable to place their work in its proper global context.

This book has been written for both international business specialists and scientists/engineers. The material contained between these covers has taken shape over several years in the "laboratory" of a graduate-level classroom at The George Washington University business school, for a course titled International Science and Technology. This course serves international business students and students who are part of the business school's Program on Science, Technology, and Innovation, of which I am director. About half the students taking the course have had scientific or engineering training and half have not. Happily, the diversity in student backgrounds makes for a stimulating classroom experience rather than for confusion and poor communication. Experience in teaching the course has shown me that, with care, it is indeed possible to bridge the gap between the two cultures.

Numerous people were helpful to me while I was writing this book. I would like to give special thanks to Jack McCarthy and Eric Winslow, chairmen of the Management Science Department during the past several years. Each graciously offered me support in the form of research and clerical help. I would also like to offer special thanks to Phil Grub of the International Business Department, whose initiative made possible the close interaction between the International Business Program and the Program on Science, Technology, and Innovation. Many students contributed to this work, both directly and indirectly, and they should be thanked. Most helpful were Debby Del Mar, Laslo Gross, Shahriar Shahida, and Carolyn Stettner. Finally, I would like to single out my wife Nancy, an international lawyer, as deserving special thanks. She provided me with a continuous stream of material in the areas of international law and trade that proved very useful to me. She also kept my two-year-old, Katherine, from pulling the plug of my word processor at some very crucial moments.

Introduction: Technology and International Business

The American Challenge

Toward the end of the 1960s, a provocative book appeared titled *The American Challenge*. In this work, the French writer Jean-Jacques Servan-Schreiber declared that European business lagged behind U.S. business because of U.S. superiority in management skills and technology. The issues raised by Servan-Schreiber were debated intensely on both sides of the Atlantic in universities, business circles, and governments. A number of studies were commissioned that examined the extent to which a technological gap existed between the Americans and the Europeans.

Today we have the advantage of seeing, through historical hindsight, that even as *The American Challenge* was going to press, the seemingly unstoppable American technological and economic juggernaut was losing steam. This fact became ever more obvious with the progression of the 1970s.

The decade of the 1970s was a traumatic one for the United States. At the opening of the decade, the U.S. economy and society were being buffeted by an endless war in Southeast Asia. This conflict, costing the United States some $30 billion a year in direct expenses, was a terrible drain on the economy. Worse still, it led to a rending of the American social fabric that would have negative consequences long after the Vietnam war was over. Civil disobedience was commonplace. Respect for many public institutions disappeared. This situation was further exacerbated by the Watergate scandal, which saw the president of the country driven from office in shame and a number of the highest public servants in the land sent to prison on charges of obstructing justice.

Meanwhile, a new international milieu was developing in which the Americans had difficulty functioning. Most notably, in 1973-1974 the U.S. economy received a devastating blow from a wholly unanticipated source: the oil-producing countries of the Third World. The rapid quadrupling of crude-oil prices, and the inability of the world's mightiest country to do anything about it, had both economic and psychological consequences. For one thing, spiraling energy costs became a significant contributor to inflation in the U.S. economy. For another, the United States found that for all of its technological might, planning skills, and managerial ability, it was essentially unprepared to deal with the energy shortages induced by the oil crisis. Little attempt had been devoted to developing energy alternatives

1

to oil, except for some attention that had been focused on nuclear energy; and the nuclear option was increasingly unacceptable to the American public. With the oil embargo, Americans found themselves dancing to tunes played by a cartel of oil-producing countries that felt little sympathy for the plight of oil consumers.

The energy crisis did more than create discomfort for the United States. It made it evident that there were some fundamental problems with the American economic system. This is most clear when we examine the responses of some other countries to the cartel-induced energy shortages. The most dramatic example is Japan, a country that depended heavily on foreign oil to meet most of its energy needs. Japanese industrial production declined in the immediate aftermath of the oil embargo. However, in a very short time, production climbed to pre-embargo levels. Similarly, the industrial performance of some European countries and even small countries, such as Korea, was quite strong in the 1970s, despite their dependence on foreign oil supplies.

What we see in the United States in this period is quite different. U.S. products faltered seriously in world markets. What is worse, many American industries were unable to compete adequately with foreign firms on their own turf: in the 1970s, foreign companies made tremendous inroads into the American market, most notably in crucial industries such as steel, textiles, footwear, consumer electronics, and automotive vehicles. Consequently, the U.S. suffered uncharacteristic balance-of-trade deficits in these areas. These deficits, coupled with unchecked inflation and periodic recession, caused the dollar, once the cornerstone of the international financial system, to weaken dramatically.

Many explanations have been offered as to why the United States slid so rapidly from being the economic behemoth portrayed in *The American Challenge* to being the hamstrung, bumbling, do-no-right giant of the seventies. Explanations focus on inept leadership; the costs of simultaneously waging a war in Vietnam and a domestic war on poverty; international circumstances, including the oil crisis; and lack of adequate levels of new investment in American industry. This last explanation received a great deal of attention as it became clear that America's economic crisis was substantially tied to its lack of competitiveness in both global and domestic markets. America's lack of competitiveness was often attributed to its aging industrial plant, a consequence of low levels of capital invested by industry. But in more recent times, a consensus began building in the United States that America's competitive problems go beyond this, that the American business crisis was largley rooted in managerial shortcomings. The basic criticisms of U.S. management are:

1. *It focuses principally on financial transactions.* For example, corporate growth and survival strategies are often more concerned with fi-

nancially driven mergers than with developing new assets from scratch. If a firm desires to develop new capabilities, it shops around with a view to buying these capabilities ready-made rather than developing them internally. The emphasis on finance also leads firms to adopt a short-term outlook. For one thing, in an inflationary period financial tools—such as internal rate-of-return analyses—make long-term investments appear unattractive. For another, in equity based companies, a financial approach tends to avoid risk and to encourage non-risky, short-term gains so that the firm's quarterly reports look good to stockholders. Finally, in an inflationary period, profits tend to be overblown and lead financial officers to have an exaggerated sense of their accomplishments.

The financial emphasis in American firms is also reflected in the fact that in recent times the fast track to the top of the corporate hierarchy has been through the financial department, as opposed to the production or research and development (R&D) departments.

2. *U.S. management does not pay sufficient attention to production.* In part, this is a consequence of the changing structure of the U.S. economy. Not so long ago, industries manufacturing tangible goods (including agricultural products) employed nearly two-thirds of the American workforce, while services accounted for somewhat more than a third. Today the figures are reversed. The U.S. economy is overwhelmingly a service economy. This shift in the composition of the economy has resulted in a corresponding deemphasis by management on production. Until recently, few American graduate schools of business offered more than the most cursory treatment of the production function. It is questionable whether more than a small fraction of the graduates of these schools have ever seen the inside of a factory. Yet maintaining competitiveness requires state-of-the-art production capabilities.

3. *U.S. management pays insufficient attention to research and development.* R&D is inherently a risky undertaking. The more long term the R&D, the greater the risk. However, truly dramatic breakthroughs in innovation are often associated with long term R&D efforts.

In large measure, the strength of the American economy in the post-war era was a consequence of the enormous R&D efforts of the 1940s. Significant innovations resulting from this R&D—for example, the discoveries of radar, operations research, the transistor, jet aircraft, atomic energy—helped carry the United States through the profitable 1950s and 1960s. In the 1960s, both industrial and government attention to R&D began to wane. R&D expenditures as a percentage of sales and gross national product (GNP) declined. In the late 1970s, a consensus emerged in both industry and government that innovation was slipping in the United States, and that this was a direct result of the R&D cutbacks of the previous years. The very end of the decade witnessed an increase in R&D spending that continued into the 1980s.

4. *U.S. management is unprepared to deal with the new international environment.* U.S. management—particularly management in large U.S.-based multinational corporations—was accustomed to pursuing its overseas business through subsidiaries. The subsidiaries were, typically, direct extensions of the home office; control over their activities was substantial. However, a changing international business environment, with its attendant dilution of American economic clout, required multinationals to deal increasingly with unaffiliated foreign firms. As foreign markets became more important (in 1960, only 5 percent of the U.S. GNP was dedicated to exports; by 1980, 13 percent), competition from foreign sources became more intense.

American management was ill-prepared to deal with the new international environment. Its knowledge of foreign languages, foreign customs, international relations, and the intricacies of international trade was minimal. These deficiencies have hurt American business in foreign markets. While management has recognized and begun, through training programs and hiring policies, to rectify some of these inadequacies, it will be many years before they are overcome.

Aims of This Book

This very brief discussion of America's problems in the 1970s serves to illustrate some of the difficulties of managing an economy in our complex, heavily interdependent world. The fortunes of even the richest and technologically most advanced countries are not assured. To a large degree, the very technology that provides a country with its advantages may—if that technology becomes obsolete or ill-suited to domestic conditions—lead to its undoing.

This book will not focus on America's economic woes. Indeed, many of the American problems of the 1970s began to fade in the 1980s, as U.S. businesses and government consciously strove to overcome some of the more egregious transgressions of the past. Instead, this book will attempt to give the reader a better appreciation of how the economic health of nations (including the United States) is often closely tied to their scientific and technological capabilities.

Our principal aim is to provide an overview of the scientific and technological environment in which international business is conducted. This environment is complex and multidimensional.

There is a *legal dimension.* Scientific and technological activities on the international level are governed by a vast web of international and domestic laws. These laws deal with all aspects of international science and technology, from defining scientific and technological responsibility to determining how international technological transactions can be carried out.

There is an *economic dimension*. Science and technology can play a significant role in the development of rich, poor, and in-between countries. They can also give countries a competitive edge that strengthens their balance-of-trade positions.

There is a *political dimension*. Science and technology can be used to serve foreign-policy objectives. In addition, they frequently appear as important agenda items in international forums, where the less-developed countries demand access to the benefits of scientific and technological developments.

There is a *military dimension*. International high-technology arms transfers are big business. Furthermore, the demands of military procurement frequently lead to dramatic new technologies whose value extends beyond their military applications.

There is a *social dimension*. Rapid change is a fellow-traveller of science and technology. The ability of a society to adapt to rapid change will help determine how successfully it meets its citizens' needs and survives in a relentlessly hostile world.

There is an *administrative dimension*. Science and technology do not just happen. They are the consequence of large numbers of people planning, funding, and carrying out scientific and technological activities.

Finally, science and technology have a *historical dimension*. They have evolved over the centuries to their present states. In some areas of the world their development was very sophisticated, while in others it was trivial. Even in the most advanced countries, their development has reflected broad historical currents, with science and technology spurting forward in times of war and stagnating in times of economic travail.

We will examine each of these dimensions in detail. We hope that this analysis will lead to a better understanding of the extent to which international business is driven by scientific and technological developments.

Definitions

In the public mind, the distinction between science and technology is vague, if it is perceived at all. Frequently, the popular press will marvel at the miracles of modern science, where science is taken to include activities that range from landing a man on the moon to putting the MFP into Colgate dental cream to gene splicing. Even among specialists, it is not always clear what the distinctions are between science and technology. It is generally agreed that there is a difference between these two realms of activity, but the precise nature of this difference is not always easy to discern. In general, it is held that science is directed toward knowledge, while technology is directed toward use. But this distinction is still not helpful in enabling us to determine, for example, the extent to which Marconi's work on the wireless was essentially a scientific or technological activity.

For most purposes, the terms science and technology are not interchangeable. First, it is important to note that these activities are usually carried out by different groups of people: science by scientists, trained in such disciplines as molecular biology, theoretical physics, and organic chemistry; and technology by engineers, trained in disciplines such as mechanical engineering, electrical engineering, and chemical engineering. Second, these two groups go about their research in different ways: scientists employ rigorous controls on their experiments, while technologists loosen these controls considerably. Third, these two groups take radically different approaches to disseminating their findings: scientists publish their findings in the open scientific literature, while technologists, recognizing that the knowledge they possess has commercial value, keep their findings to themselves. When a technologist makes his knowledge public, it is likely to be in the form of a patent. Finally, the principal rewards of scientists and technologists are different: the typical scientist is most interested in obtaining recognition from his peers, while the typical engineer is more interested in material rewards, or in obtaining satisfaction from the realization that he has produced something useful.

To complicate matters further, scientific research is often broken down into two categories: basic research and applied research. Basic research—sometimes called pure research—is undertaken with no particular end use in mind. It involves the pursuit of knowledge for the sake of knowledge. Applied research entails the acquisition of knowledge that is directed toward some ultimate end use.

Once again, the definitional distinctions are not always very clear in practice. In their discovery of the double helix structure of the DNA molecule, Crick and Watson paid little attention to how this knowledge could be practically employed. The discovery of the refrigerant Freon by Frigidaire scientists exemplifies, par excellence, applied science in the service of industry. The invention of the zipper by Judson and Sundback is a purely technological accomplishment, based on no scientific principles. In each of these cases, there is little ambiguity as to whether we are talking about basic science, applied science, or technology. However, in many cases the distinctions are blurred. Is the bioscientist who is attempting to find a cure for cancer by means of genetic research engaging in basic or applied research? He is motivated in his research by patently practical concerns (for example, to save lives), yet his explorations touch upon some of the most fundamental questions of the nature of life.

The term *research and development* also leads to some confusion. Of course, this is not really one term but two. However, when research and development are formulated as *R&D*, they are commonly treated as a single entity. Research is directed toward the acquisition of knowledge. Development entails the translation of this knowledge into something that is tangible and and—frequently—that possesses commercial value. The important thing to note about the relationship of research to development is that research generally consumes far fewer resources than development. Typically, for every dollar spent on research, five to ten dollars is spent on development. The reason for this difference in resource consumption is that research can usually be carried out on a small scale, employing relatively inexpensive equipment. Development, however, is generally carried out on a much larger scale and may involve the building of a pilot plant, the purchase of buildings, the tailor-made construction of unique full-scale equipment, and so forth.

In this book, we will not concern ourselves with definitional niceties. Suffice it to say that science and technology are neither synonymous nor always perfectly distinct. We offer these definitions of terms.

science: the systematic acquisition of knowledge about how things work in nature

technology: man's employment of knowledge to extend control over his tangible environment

Science, Technology, and Economic Development

Even a rudimentary reading of modern Western history shows that scientific and technological advances have had momentous economic repercussions. Indeed, the development of the West closely parallels technological advances. Today, the most economically advanced countries are those with strong scientific and technological capabilities. Technology can be called the engine of economic growth.

The impact of science and technology on the development of society is pervasive. Some consequences of scientific and technological activity are immediate and obvious, while others are more indirect. McCormick's reaper, for example, had rather straightforward implications. This machine helped revolutionize agriculture by mechanizing the harvesting process. One machine could do the work of several men laboring with scythes. Agricultural productivity began its steep ascent. Manpower was released to work in industrial centers. Increased productivity assured ample food supplies at reasonable prices. Ultimately, technological advances in agriculture would enable 3.1 percent of the U.S. workforce to account for 6.1 percent of national income.

Many scientific and technological advances have had a less direct economic impact, though they were still of enormous consequence. Medical advances, for example, have helped economic development by creating an environment amenable to high productivity. Much of the great human and economic waste associated with traditional scourges—such as tuberculosis, plague, polio, and small pox—has been largely eliminated as a result of advances in medicine and public health.

While it may seem that there is an undeniable relationship among science, technology, and economic development, the precise nature of this relationship is exceedingly complex and poorly understood. The common wisdom of our day would posit the relationship symbolically in the following way:

$$\text{science} \longrightarrow \text{technology} \longrightarrow \text{productivity} \longrightarrow \text{economic growth}$$

where the arrows can be read as *leads to*. This view asserts that scientific research, which enables us to have a better understanding of nature, is translated by scientists and engineers into useful applications in the form of technology. One of the principal features of technology is that it enables us to do something better than we could before its advent. Often this means that we get more output per unit of input, which is to say that technology leads to increased productivity. With continuous increases in productivity, we find that a fixed number of laborers are able to produce more and more goods and services. This is tantamount to saying that per capita wealth is increasing—in short, increased productivity yields economic growth.

This sequence of events does indeed describe some situations. For example, basic research undertaken at Bell Laboratories in the 1940s resulted in the development of the transistor. After a number of years, the transistor was employed for practical ends, first by the military, then by the industrial sector, and ultimately for popular consumption. It evolved in complexity from a mere substitute for vacuum tubes to a highly complicated computer-on-a-chip (microprocessor). Its employment in countless ways (in computers, telecommunications, office machinery) has contributed substantially to Western productivity increases and economic development.

Despite the fact that the transistor conveniently fits the causal sequence posited earlier, it would be inaccurate to generalize from this and other obvious examples. For one thing, the very first link in the chain—that is, science→technology—does not always hold. Although some historians of science and technology would dispute this, most maintain that technology has often developed independently of science. Many of the great inventions of the past few centuries were made by men with little or no scientific training. In fact, the causal sequence was occasionally opposite of what has been posited here. Numerous scientific discoveries had to await technological advances made by craftsmen, particularly in instrumentation. On occasion, technological developments stimulated major scientific breakthroughs. For example, the development of the high pressure steam engine led to major advances in the formulation of the laws of thermodynamics.[1]

The remaining links of the chain also provide us with difficulties. We know for certain that technological advances and economic growth go hand in hand. Not only is this intuitively evident, but it has also been demonstrated empirically by economists such as R. Solow and E.F. Denison.[2] However, on some occasions technology spurs economic development, while on others the economic sphere shapes technological advances. The first situation is labeled *technology push*, while the second is called *demand pull*.

The technology-push mechanism is lucidly illustrated in a passage written by Adam Smith in his *Wealth of Nations* (1776). Smith visited a small pin-making factory that employed only ten people. He observed:

> Those ten persons . . . could make among them upwards of forty-eight thousand pins a day. Each person, therefore, making a tenth part of forty-eight thousand pins, might be considered as making four thousand eight hundred pins in a day. But if they had all wrought separately and independently, and without any of them having been educated to this peculiar business, they certainly could not each of them have twenty, perhaps not one pin in a day.[3]

In this example, the effective employment of technology clearly resulted in remarkable productivity gains and contributed to economic growth.

The demand-pull mechanism finds entrepreneurs developing technologies that take advantage of market opportunities. For example, after the Arab oil embargo of 1973-1974, it became abundantly apparent that society would need to gear its energy requirements to nonpetroleum technologies. Economic exigencies led to a massive focusing of research on opportunities in solar, geothermal, gasification, liquefaction, and other nontraditional energy areas.

The existence of technology demand-pull was illustrated empirically by Jacob Schmookler for several industries.[4] Using patents as an indicator of inventive activity, Schmookler showed that in a number of industries, inventive activity consistently followed capital outlays. Thus, several years after an increase in, for example, capital expenditures in the railroad industry, there would be corresponding increases in the number of patents awarded in relevant railroad-technologies.

Technology, then, is both a shaper of and response to economic activity.

The International Product Life Cycle and the Diffusion of Technology

In the 1950s and 1960s the United States was the world's premier manufacturer of television sets. At that time, television incorporated what was, for the era, high technology. As television production became routinized and the technology moved from being state-of-the-art to being commonplace, production shifted overseas, where it could be undertaken more cheaply than in the United States. The United States now became a net importer of a product that it, more than any other country, was responsible for developing. The case of television provides us with a clear example of the functioning of the international product life cycle.

The Product Life Cycle

The idea of an international product life cycle is a natural extension of the conventional product life-cycle concept. The conventional view of the product life cycle is portrayed in figure 1-1. Four stages are identified: innovation of the product; product growth; product maturity; and product decline. The innovation stage is characterized by very low levels of sales and negative profits. These negative profits are generated by the fact that sales are low or nonexistent, coupled with the high costs of developing and launching a new product. In the growth stage, profits and sales rise quickly. Because this is a brand-new product, the firm enjoys a large share of the market. The

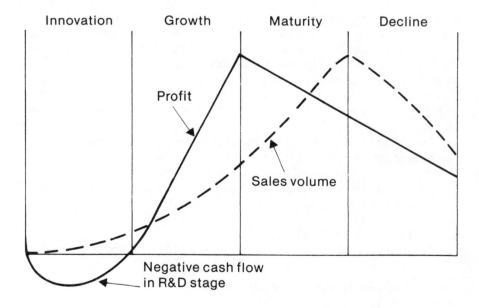

Figure 1-1. The Product Life Cycle

product becomes mature when profits peak and then steadily decline, despite the fact that sales continue to increase. The reason that profits decline is that competition for the product increases, so that its price is driven down. Finally, the product enters the stage of decline when sales also begin a downward descent. Sales generally begin to drop when the product becomes obsolete or the buying public loses interest in it.

The international product life-cycle examines the relationship between sales of a product and the diffusion of technology from an innovating country to imitating countries. One way to look at the international product life-cycle is to investigate it—as we did the conventional product life-cycle—in four stages. In stage 1, the innovating country enjoys export strength, since it is in a monopoly position in respect to the product. In stage 2, foreign firms obtain access to the technology through licensing arrangements, direct investments from the innovating country, or their own R&D efforts. With access to the technology they begin their own production of the good. In stage 3, economies of scale coupled with low labor-costs, enable the imitating country to compete effectively with the innovating country in third-country markets. Finally, in stage 4, the imitating country has costs so low that it can now export the product into the innovating country and compete effectively with domestically produced goods.

As demand for the product increases in third-country markets, those countries with even greater comparative advantage in labor begin producing the good. As soon as they can fully realize economies of scale, they can produce the good more cheaply than either the innovating country or early imitating countries. Consequently, they become major exporters of the good and compete very effectively with the innovating and early imitating countries in international markets. At this time, the innovator and early imitator may become net importers of the good.

The overall workings of the international product life-cycle and the diffusion of technology are nicely illustrated in the case of semiconductors. The following account is taken from J.E. Tilton's review of international developments in semiconductor production in the 1950s and 1960s.[5] In this period, most of the major breakthroughs in semiconductor technology originated in the United States. Many of the basic breakthroughs were funded by the government and were designed to meet defense needs. In defense procurement, government is often more concerned with product performance than with product cost, so semiconductor manufacturers were able to develop technologies that might not have been viable in the conventional marketplace. Furthermore, with a guaranteed government consumer, risks were minimized and the semiconductor researchers and manufacturers could attempt to make daring breakthroughs without fear that their adventurous ways would bankrupt them.

These semiconductor devices were produced solely in the United States and met the demands of the government-defense market. As yet, they were too expensive to be employed by the industrial or consumer sectors. However, with time and increased production, the manufacturers of semiconductor devices gained experience in production and realized economies of scale. The cost of producing the new devices decreased to the point where it became affordable to the industrial sector.

At this point, European interests in producing the new technologies heightened, since a demand existed in European industry for industrial applications of semiconductors. The American-developed technology was then transferred to the Europeans, who then became competitive producers of the new technology. Their competitiveness was due, in large measure, to their cheaper labor-costs.

With both American and European production of semiconductors, production skills improved, further economies of scale were realized, and the price of the new semiconductor devices were driven so low that the devices could be employed in consumer goods. At that time, the Japanese began producing the semiconductor devices. With learning economies, economies of scale, and wage rates substantially lower than the Europeans, they were able to produce semiconductor devices that competed very strongly with American and European devices.

Some Factors Affecting the Rate of
Diffusion of Technology

The account provided here of the international product life-cycle and the diffusion of technology is highly idealized. Our presentation makes it appear to be a well-oiled, smoothly functioning process where technologies are transferred effortlessly from country to country. In the real world there are a number of factors that clutter the lean picture painted here. Some of these factors will be discussed cursorily here, and in greater depth later.

Trade Barriers. The speed with which technology is diffused among countries is dependent to a large extent on the presence or absence of trade barriers. These barriers can take many different forms. At the broadest level, there are barriers to exports and barriers to imports. Typical of export barriers are export control laws, such as the Export Administration Act of 1979 in the United States. This particular law restricts the export of goods and technology if such exports are detrimental to U.S. national security; or if the withholding of such exports furthers U.S. foreign-policy objectives; or if the exports have a serious negative impact on the U.S. economy.

Import barriers can generally be classified as tariff or nontariff barriers. Tariffs are simply taxes imposed on imports that raise the price of the imports in the domestic market. Tariffs are most commonly imposed to protect domestic industry from foreign competition. They can serve other purposes as well. For example, a less-developed country (LDC) may impose heavy import-duties on luxury items in order to avoid conspicuous consumption.

Nontariff barriers, as the name implies, restrict imports by means other than tariffs. Import quotas, government and industry procurement policies that discriminate against foreign goods, and administrative protection measures that entail substantial red-tape for the importation of goods are examples of non-tariff barriers.

In general, the greater the trade barriers, the greater the impediments to the diffusion of technology. For example, the prohibition of all exports of a military technology will put a serious crimp on its spread (which may be just as well, if the proliferation of the technology makes the world a more dangerous place). The only way it will be diffused to different countries is if it is stolen or reinvented elsewhere. Heavy import-duties will also inhibit the diffusion of technology if they keep products out of the economic reach of domestic consumers. On the other hand, if a large domestic demand already exists for the products, heavy import-duties may encourage the exporter of the technology to enter into a licensing agreement with a domestic firm in order to capitalize on high demand while bypassing import duties. In this case, trade barriers actually accelerate the international diffusion of technology among manufacturers.

Absorptive Capacity of the Technology Recipient. The absorptive capacity of the technology recipient will strongly color the character of the international product life-cycle and the diffusion of technology. If the technology recipient has substantial scientific and technological capabilities, it can readily absorb the technology and will have little dependence on the technology donor. In this case, diffusion can occur rapidly, and the transfer of technology may be profound, giving the recipient new capabilities that will enable it to become technologically self-sufficient. If the technology recipient has weak scientific and technological capabilities, the diffusion of technology might be rather slow and the transfer superficial. Here the recipient will generally be heavily dependent upon the donor for know-how.

Patents. Technology is often protected from unauthorized copying by patents. A patent gives an inventor monopoly rights over his invention for a specified period of time. If the inventor chooses to maintain close control over his technology and refuses to share his patent with others, this can have a severe restrictive effect on the diffusion of the protected technology.

Sundry Other Factors. The international product life-cycle model presented here assumes an effortless transfer of technology from country A to country B. In practice, the transfer occurs in a number of specific ways. The developer of the technology may license the use of his technology to a foreign producer, enabling him to employ the technology on his own, subject, of course, to the terms of the licensing agreement. Or the developer of the technology may set up a subsidiary in the foreign country and produce the technology himself. In this event, he maintains direct control over the technology. Or he may undertake a joint venture with a firm in another country, exploiting the technology with a partner, and sharing in the costs of its production and distribution. The technology can also be considered transferred if a party in another country employs its indigenous R&D capabilities to duplicate the original technology, or to develop a similar technology. Generally, the diffusion of technology that takes place in the international product life-cycle takes on one of these four forms. However, as we shall see later, these forms can be supplemented by many ancillary technology-transfer mechanisms, such as on-the-job training and the selling of technical services.

Technology and the U.S. Balance of Trade

The 1970s saw the United States go from a position of consistent trade surplus to one of chronic deficit. Table 1-1 illustrates the difference in the U.S. balance of trade in 1970 and 1980. In 1970 the trade balance is barely positive; in 1980 the United States is $25 billion in the hole. To a large extent,

Table 1-1
U.S. Exports and Imports of Merchandise, 1970 and 1980
(millions of dollars)

	1970	*1980*
Total exports	41,980	223,966
Developed countries	29,447	136,915
LDCs	12,165	82,908
East Europe	368	4,143
Total imports	39,870	249,380
Developed countries	29,014	128,798
LDCs	10,638	119,138
East Europe	218	1,444
Balance of trade	2,110	− 25,414

Sources: Bureau of Economic Analysis, *Survey of Current Business,* 52 (March 1972):52-55; and 62 (March 1982):59-63.

the deficit is accounted for by the dramatic increase in U.S. importation of foreign oil, coupled with the spectacular rise in oil prices after 1973. When petroleum is removed from the trade statistics, we find the United States with a healthy trade surplus.

When the trade statistics are further disaggregated, an interesting finding emerges: the United States performs very well in R&D-intensive areas. This finding is not surprising in view of our previous discussion of the international product life-cycle. It should be recalled that at the earliest stage of the international product life-cycle, an innovating country is an exporter of technology, for the very straight-forward reason that it is the sole producer of the technology. Only when the technology has been around for a while, and imitators begin producing goods based on it, does the innovating country lose its export advantage.

The statistics in table 1-2 show how good U.S. trade performance has been in R&D-intensive areas. U.S. exports of R&D-intensive manufactured products consistently outstrip imports in 1960-1980. Furthermore, the balance of trade for R&D-intensive products has grown to very substantial proportions: from $5.9 billion in 1960 to $52.4 billion in 1980. In contrast, the trade balance in non-R&D-intensive manufactured products has been consistently negative in 1960-1980. The deficit for these products has closely paralleled—in a negative sense—the gains for R&D-intensive products: the deficit grew from $179 million in 1960 to $33.4 billion in 1980.

The encouraging trade statistics for R&D-intensive manufactured products have a dark side to them. Although the absolute value of the balance of trade for these products has grown substantially, imports have been growing more rapidly than exports: between 1960 and 1980, exports grew thirteen-fold while imports grew twenty-seven-fold. If such a trend should con-

Table 1-2
U.S. Trade Balance in R&D-Intensive and Non-R&D-Intensive Manufactured-Product Groups, 1960-1980
(millions of dollars)

	R&D-Intensive			Non-R&D-Intensive		
	Balance	*Export*	*Import*	*Balance*	*Export*	*Import*
1960	5,891	7,597	1,706	− 179	4,962	5,141
1965	8,148	11,078	2,930	− 2,027	6,281	8,308
1970	11,722	19,274	7,552	− 8,285	10,069	18,354
1975	29,344	46,439	17,095	− 9,474	24,511	33,985
1980	52,384	98,324	45,940	− 33,444	45,646	79,090

Sources: National Science Board, *Science Indicators 1980* (Washington, D.C.: U.S. Government Printing Office, 1981), p. 234; and U.S. Department of Commerce, *Overseas Business Reports: U.S. Foreign Trade Annual, 1974-1980* (Washington, D.C.: U.S. Government Printing Office, November 1981).

tinue indefinitely, imports of R&D-intensive manufactured products might match and then exceed exports sometime in the not-too-distant future.

Table 1-3 shows U.S. export and import performance for a select number of R&D-intensive manufactured-product areas. Although not seen in this table, the most dramatic gains in the balance of payments have been in the category of nonelectrical machinery. The most notable contributions to this category came from increased exports of electronic computers, internal combustion engines, construction equipment, and mining and well-drilling machinery.

When viewed from the perspective of international markets, the greatest increases in the U.S. trade balance for R&D-intensive manufactured products have been with LDCs. The LDCs purchase one-third more of these products from the U.S. than the West Europeans. Furthermore, the positive trade balance for these products is twice as great with the LDCs as with West Europe.

Table 1-3
Trade Balance in Selected R&D-Intensive Manufactured-Product Groups, 1960-1980
(millions of dollars)

Product Groups	1960	1965	1970	1975	1980
Chemicals[a]					
Balance	955	1,634	2,376	4,995	12,157
Export	1,776	2,403	3,826	8,691	20,740
Import	821	769	1,450	3,696	8,583
Machinery[b]					
Balance	3,752	5,135	6,311	17,245	24,977
Export	4,476	7,935	11,685	29,215	57,263
Import	724	1,800	5,374	11,970	32,286
Aircraft					
Balance	970	990	2,382	5,617	10,931
Export	1,024	1,130	2,656	6,136	12,816
Imports	54	140	274	519	1,885
Professional & scientific instruments					
Balance	214	389	653	1,487	7,319
Export	321	610	1,107	2,397	7,505
Import	107	221	454	910	186

Sources: National Science Board, *Science Indicators 1980* (Washington, D.C.: U.S. Government Printing Office, 1981), p. 235; and U.S. Department of Commerce, *Overseas Business Reports: U.S. Foreign Trade Annual, 1974-1980* (Washington, D.C.: U.S. Government Printing Office, November 1981).

[a]Includes drugs.
[b]Includes electrical and nonelectrical, computers, and communications equipment.

In contrast, the United States suffered a negative trade balance with Japan for R&D-intensive manufactured products over a period of some two decades. Until the mid-1970s, the negative balance stood in the range of several hundred million dollars per year. In the mid-1970s, it suddenly jumped, and by the end of the decade hovered in the $4-5 billion range.

Some concern has been expressed by knowledgeable American observers of international technological developments that firms in the United States engage in technology management policies that negatively affect the competitive position of U.S. business in world markets. The argument is that the United States contributes far more to the world pool of know-how than it withdraws. On the one hand, it is argued, American firms part with their technological resources too readily, and in effect give away the fruits of their R&D labors. On the other hand, they ignore technological developments overseas that might be profitable to them.

The first argument—that U.S. firms engage in technology give-aways— is illustrated in the case of Amdahl Corporation's development of an advanced computer that would compete head-on with IBM's most sophisticated machines.[6] In the early 1970s, Amdahl Corporation had difficulty raising cash from American sources for its venture. The company obtained substantial financing from the Japanese computer manufacturer, Fujitsu Ltd. Soon after receiving this support, Amdahl and Fujitsu entered into a royalty-free cross-licensing agreement that gave each firm access to technology being developed by the other. This agreement gave Fujitsu the ability to incorporate Large Scale Integrated Circuit (LSI) technology into its computers. At the time LSI technology was unique and the Japanese had not yet made inroads into successfully developing it themselves. In addition to obtaining much-needed financing, Amdahl also received production-engineering and technical support from Fujitsu.

When examining technology transfer in this case, it is evident that the flow of unique technology was one-directional—from Amdahl to Fujitsu. While Amdahl benefitted from this arrangement, it is likely that the overall U.S. computer industry suffered, because it helped strengthen the competitive position of a Japanese firm. Incidentally, it is interesting to note that the Amdahl-Fujitsu collaboration was not unique. In the 1960s, all the major Japanese computer companies—Hitachi, Mitsubishi, Nippon Electric, Oki, and Toshiba—entered into licensing arrangements with U.S. firms—RCA, TRW, Honeywell, Univac, and GE respectively. One reason why the U.S. firms entered into these agreements was that Japanese law restricted the ability of foreign firms to manufacture computers in Japan.[7]

The argument that other countries benefit handsomely from U.S. technology also has statistical support. Table 1-4 presents data on U.S. receipts and payments of royalties and fees for unaffiliated operations (it excludes royalties and fees related to foreign direct investments). These data are a

Table 1-4
U.S. Receipts and Payments of Royalties and Fees Associated with Unaffiliated Foreign Residents, 1970 and 1980
(millions of dollars)

	1970	1980
Total net receipts	573	1,170
Developed countries (less Japan)	307	603
Japan	202	354
LDCs	65	213
Total net payments	114	255
Developed countries (less Japan)	104	225
Japan	4	19
LDCs	7	11

Sources: National Science Board, *Science Indicators 1980* (Washington, D.C.: U.S. Government Printing Office, 1981), p. 231; and Bureau of Economic Analysis, *Survey of Current Business,* 62 (March 1982), pp. 59-63.

rough indicator of U.S. technology licensing activity with unaffiliated foreign operations. They suggest that U.S. firms license-out far more technology than they license-in. For example, in 1980, some $1,170 million in royalties and fees were received by American firms from unaffiliated foreign operations, while only $255 million was paid out. Particular attention should focus on receipts from and payments to Japan. In 1980, the ratio of receipts to payments was nineteen to one! Since these payments and receipts are largely for the acquisition of technology, these data suggest that Japan is receiving far more U.S. technology than it is offering to the United States.

The data in table 1-4 also support the second argument previously mentioned—that U.S. firms pay insufficient attention to foreign R&D efforts. This habit of ignoring foreign works was ingrained in the immediate postwar era. At that time, it could be reasonably said that the overwhelming bulk of the world's significant R&D activity occurred in the United States. There was little reason to look outside the confines of the United States for R&D ideas and results. Through the years, U.S. industry and the U.S. R&D community held to this view, even after foreign R&D capabilities strengthened. Observers who became concerned by American ethnocentrism in this respect maintained that the United States was suffering from a not invented here (NIH) syndrome. It is impossible to assign a dollar value to opportunities lost to the United States on account of the NIH syndrome, but simple logic suggests that the figure must be substantial.

The NIH syndrome seems to hold in the basic and applied sciences as well as in technological areas. A number of studies have shown that, in all fields of science, foreign scientists refer (through footnotes) to American works far more frequently than American scientists refer to foreign works.

This finding suggests that far more information is flowing from the United States to the rest of the world than from the rest of the world to the United States.[8]

Conclusions

Science and technology permeate the international business environment. A country's strength in global markets is dependent to a large extent on its technological capabilities. This is a major reason why Third World and Eastern Bloc countries are so eager to obtain advanced Western technology. They are painfully aware that without it they have difficulty producing goods that are competitive—in both price and performance—in the world market place.

In the next two chapters, we continue our macroscopic view of the world, and examine scientific and technological efforts in developed and less-developed countries. Thereafter, we shift to a more microscopic outlook and examine in greater detail how technological resources are mobilized in international business.

Notes

1. J. Jewkes, D. Sawers, and R. Stillerman, *The Sources of Invention* (New York: W.W. Norton & Co., Inc., 1969), p. 64.

2. The seminal works that launched this avenue of investigation were R. Solow, "Technical Change and the Aggregate Production Function," *Review of Economics and Statistics* 29 (1957):312-320; and E.F. Denison, *The Sources of Economic Growth in the United States* (New York: Committee on Economic Development, 1962).

3. A. Smith, *The Wealth of Nations* (New York: The Modern Library, 1937), p. 3ff.

4. J. Schmookler, *Invention and Economic Growth* (Cambridge: Harvard University Press, 1967).

5. J.E. Tilton, *International Diffusion of Technology: The Case of Semiconductors* (Washington, D.C.: The Brookings Institution, 1971).

6. J. Baranson, *Technology and the Multinationals* (Lexington, Mass.: Lexington Books, D.C. Heath and Company, 1978), pp. 75-83.

7. Ibid., p. 74.

8. See, for example, J.D. Frame, "Cross-National Information Flows in Basic Research: Examples Taken from Physics," *Journal of the American Society for Information Science* 29 (September 1978):247-252.

2

Science and Technology in Advanced Countries

Introduction

At a superficial level what distinguishes the developed countries (DCs) from the LDCs is that the former have higher per capita incomes than the latter. That is, they are richer than the LDCs. An examination of some countries with newly acquired wealth—such as Saudi Arabia, Kuwait, and Bahrain—causes us to reflect a moment on this distinction between the DCs and LDCs. In recent years, these Arabian-peninsula countries have acquired great wealth owing to their possession of significant oil reserves and to tremendous cartel-induced price hikes for oil. Yet, even with their high per capita incomes they are not viewed as developed. The reason for this is that, despite their great wealth, they remain technologically underdeveloped. Until such time as they have a large cadre of scientists, technologists, and technicians; have high literacy rates in the overall population; and are capable of maintaining a modern industrial system without dependence on outside technical assistance, they will be viewed as less than developed.

A more meaningful distinction between DCs and LDCs would seem to be based on their differences in scientific and technological capabilities. The DCs possess such capabilities in abundance, the LDCs lack them.

It is no accident that the developed countries of the world—that is, the countries of Europe and other countries settled and populated by Europeans—have good scientific and technological abilities. These abilities were carefully nurtured over several centuries. Modern science is very clearly a child of Europe. It was the Europeans who first took a systematic approach to examining nature, who justified their speculations on the physical nature of the world through observation of physical phenomena, and who established the scientific method, whereby hypotheses are stated and tested.

While contemporary science is clearly rooted in modern Europe, the origins of technology are much more antiquated and exotic. If technology entails the use of tools by men to achieve some desired end, then its origins extend back some two and a half million years to the hominid *homo habilis,* who, while situated in the eastern and southern portions of the African continent, was the first creature on earth to fashion and employ tools consciously. *Homo sapiens,* who came after *homo habilis,* continued making and using tools, working initially with stone, wood, and bone. The develop-

21

ment of tools—and with them technology—proceeded very slowly. Man depended on stone, wood, and bones as the basic material for tools for roughly two million years. It was only between 6,000-4,000 B.C. that he began using metals. Progress in tool fabrication and use continued to proceed slowly, notable progress being measureable in millenia. Even in the comparatively modern European Middle Ages, most aspects of daily life were not much different than they had been in Roman times.

The development of technology was proceeding in all regions of the globe inhabited by man. However, the rate of technological advance varied widely from region to region. The type and sophistication of the technology varied with the environmental demands placed on man, as well as with the nature of his social organization.

While the development of technology is ubiquitous, the advanced technology of today—like science—is principally a product of Europeans and is closely tied to the industrial revolution. Beginning with the seventeenth and eighteenth centuries, technology began an explosive growth in Europe. Technological change occurred at faster and faster rates. More change could be seen in a matter of decades than in all previous human history. Society and the economy were restructured to accomodate the needs and demands of this burgeoning technology. The pastoral life gave way to the mushrooming factories and cities. Relatively independent rural workers flooded the cities in search of employment and lives as wage slaves, or proletariat.[1]

In the mid-nineteenth century Karl Marx saw these developments as creating severe tensions in the industrializing capitalist societies. In his view, the disenfranchised and oppressed proletariat, mercilessly exploited by the capitalists, would ultimately overthrow the system to create a classless society where the means of production were owned by all. What Marx did not reckon on was that the same technology that made workers disenfranchised wage slaves would ultimately improve their lives tremendously. Thanks to technological advances, it can reasonably be said that the average European assembly-line worker today lives a more comfortable and secure life than did a king in the Middle Ages. Not surprisingly, given the benefits of technology, there has not been a single Marxist-style proletarian revolution anywhere.

One of the chief features of science and technology (S&T) in the advanced countries today is the degree to which they are integrated into society. It is impossible to give an accurate portrayal of countries such as the United States, France, and West Germany without mentioning either directly or by implication the important roles played by S&T in nearly all facets of life. The economies of these countries are clearly driven by technology. Commonplace items such as pens, paper, shirts, buttons, and the like are manufactured in ways undreamed of a century ago. They are often made of new synthetic substances that have been discovered only recently (by chemists, metallurgists, materials scientists), and the production process often employs equipment that is continuously being improved, and is increasingly moving in the direction of full automation. Nowhere is the pervasiveness of technology seen more clearly than in the Sunday want-ads section of a major urban newspaper. A brief scan of the want-ads shows a demand for word-processing specialists, computer operators, systems analysts, engineers of every sort, and numerous other workers with skills that are rooted in new technologies.

The political activities of the advanced countries are also greatly influenced by technology. Countless articles have appeared in the Western press describing how politicians aim their pitches to the electronic and print media rather than directly to the electorate. Scientific polling of voter preferences has become so refined that experts can predict the outcomes of elections with great accuracy long before the polls have closed and the last ballot has been cast.

Finally, the social systems of these countries are largely reflections of advanced technology. Works such as Daniel Bell's *The Making of Post Industrial Society,* Alvin Toffler's *Future Shock* and *The Third Wave,* Jacques Ellul's *Technology and Society,* and Gary Gappert's *Post-Affluent America* direct most of their attention to the impact of technology on the social system. Some of the more visible technology-based characteristics of social systems of developed countries include: a high degree of mobility for individuals; rapid obsolescence of recently acquired skills; a nuclear family structure; a strong commitment to continuing education after formal education has been completed; substantial quantities of leisure time; and high levels of affluence.

Level of S&T Activity in Different Countries

While all the developed countries have rather substantial scientific and technological capabilities, the commitment to S&T varies notably from country to country. This is understandable in view of the different levels of human and economic resources in these countries. It has been shown that for developed countries there is a strong correlation between scientific effort and the size of a nation's economy and its level of affluence.[2] That is, a country with a large economy (measured by GNP) and a high level of affluence (measured by GNP per capita) will generally be very active scientifically, while a country with a small economy and moderate level of affluence will have a far more modest level of such activity. In between these two extreme positions, it appears that, given two countries of equal economic size (or equal levels of affluence), the more affluent (or economically larger) country is generally more active scientifically.

The explanation of this finding is fairly straightforward. The importance of economic size is obvious. At the outset of the 1980s the United States devoted roughly 2 percent of its GNP to R&D. This amounted to more than $60 billion, a figure that is roughly the same as the total GNP of Denmark. Clearly, small Denmark cannot expect to dedicate more funds to R&D efforts than its entire GNP. There is an upper limit on the resources the Danes can earmark for R&D that is determined by the small size of their economy.

The importance of affluence is somewhat more conjectural. The results seem to suggest that the greater the level of affluence, the more a country can afford to engage in luxurious undertakings.

The different levels of S&T in the advanced countries can be seen by looking at various R&D indicators. The two commonest indicators—which look at R&D manpower and R&D funding—are called input indicators, because they depict the resources that go into the R&D system. Output indicators are scarce, because it is difficult to identify quantitatively the outputs of the research system. In some cases, the output is simply an idea, in others it is a marketable product, in still others it is an improved production process, and so on. We will examine an output indicator that in recent years has gained acceptance as a valid indicator of scientific effort. This indicator is the number of research papers appearing in high quality scientific journals. Currently, there are no indicators of technological effort that are viewed as being as reliable as the publication indicators. Great interest has focused on patent statistics as possibly good technology-indicators, but there are a number of serious problems that restrict their validity as indicators. For example, the nature of the patent systems in existence throughout the world differ markedly from country to country, so that cross-national comparisons of inventive activity based on patents are difficult to make.

Table 2-1 provides input and output statistics for the world's six most significant scientific powers. These six countries together account for between 80 to 90 percent of all scientific and technological activities in the world today.

The manpower statistics in table 2-1 show that the Soviet Union has by far the largest number of scientists and engineers. The numbers given in the table are full-time equivalent (FTE) figures. Such figures are commonly employed in R&D manpower counts. Inasmuch as many scientists and engineers move into administrative positions as they proceed along their career paths, and because much of the time of practicing scientists and engineers is devoted to tasks other than research (for example, teaching and administration), FTE statistics were developed to factor out the non-R&D efforts of scientists and engineers.

Following the Soviet Union in S&T manpower is the United States, which has half the number of FTE scientists and engineers of the Soviets.

Table 2-1

International Comparison of Some Basic Indicators of Scientific and Technological Activity, Late 1970s

	Scientists & Engineers[a] (thousands)	Total R&D (percent of GNP)	Civilian R&D (percent of GNP)	Scientific Papers published (thousands)	Number of Nobel Prize Winners, 1955-1980
France	68	1.76	1.35	15	6
Japan	273	1.93	1.87	14	2
United Kingdom	80-90	2.11	1.47	25	25
United States	621	2.25	1.57	104	83
Soviet Union	1,300	3.44	0.7-2.0[b]	24	8
West Germany	111	2.36	2.18	16	8

Sources: Manpower and funding data from National Science Board, *Science Indicators 1980* (Washington, D.C.: U.S. Government Printing Office, 1981).

[a]Full-time equivalent.

[b]Estimates of Soviet R&D dedicated to defense vary from 40-80 percent. See D. Holloway, "Soviet Military R&D: Managing the Research-Production Cycle," in J.R. Thomas and U.M. Kruse-Vaucienne (eds.), *Soviet Science and Technology* (Washington, D.C.: Published for the National Science Foundation by George Washington University, 1977), pp. 189-228.

The gap between the U.S. and Soviet manpower counts is not as great as it appears from these numbers. Some of the gap is explained by definitional problems. Many of the individuals treated as scientists and engineers in the Soviet count would probably be classified as technicians in the United States.

The number of FTE Japanese scientists and engineers is half the number for the United States. In third place, West Germany has less than half the Japanese number of FTE scientists and engineers. The U.K. and France are at the bottom of the list. Interestingly, the number of scientists and engineers situated in the three big Western European countries is roughly the same as for Japan alone. In view of these statistics, Japan's ascendancy in technological areas is hardly surprising.

These manpower data should be interpreted with caution. For one thing, they are primarily body counts. They tell us nothing about the quality of scientific and technological workers. To illustrate the pitfalls of relying too heavily on manpower counts for assessing a country's scientific and technological capabilities, consider that by Unesco figures Egypt has roughly 250,000 scientists and engineers in contrast to Israel's 30,000 (note: these are not FTE counts). Yet, anyone familiar with Middle Eastern science and technology would agree that the Israeli scientific and technological effort is vastly superior to that of Egypt.

For another thing, manpower indicators, such as those presented here, tend to be highly aggregated. We cannot be certain from these aggregated numbers of the actual mix of the manpower pool. How many individuals are scientists, how many engineers? How many of the scientists are in the life sciences? Physical sciences? What proportion of the engineers are civil engineers? Mechanical engineers? Chemical engineers? and so forth. It would be nice if data that were more disaggregated were available. But the collection of such data is very expensive and frequently of dubious reliability. When disaggregated data do exist for a number of countries, different national definitions of scientists and engineers do not allow for reasonable international comparisons.

Despite their limitations, manpower data give us a rough feeling for the number of individuals working in the areas of science and technology in different countries. This helps give us a crude idea of different levels of national commitment to R&D.

When we make R&D funding comparisons for different countries, we typically look at the percentage of GNP dedicated to R&D. Obviously, for many purposes it would be more useful to compare absolute levels of R&D expenditures, to say, for example, that country A spent one billion dollars on R&D in a given year while country B spent two billion dollars. However, since R&D expenditures are measured in local currencies, we run into trouble when we try to express these varying currencies in a single unit such as the U.S. dollar. Typically, currency conversions are based on foreign exchange rates, but these exchange rates have a number of drawbacks to them, two of which are particularly significant.

First, in an age of floating rates, we find that currencies can move dramatically in respect to each other. It is not uncommon to find one currency being depreciated by 20-30 percent against another currency in a matter of several months. Comparisons of R&D expenditures, therefore, can yield dramatically different results, depending upon the state of foreign exchange at the time of the comparisons.

Second, foreign exchange rates are determined by international market forces, which may or may not reflect the value of R&D in different countries. If we wished to make reasonable international comparisons of absolute levels of R&D expenditures, we would have to create a special foreign exchange rate just for R&D activities.

We get by the problem of converting R&D expenditures to a common unit of money by looking at R&D expenses as a percent of GNP. These percentages are given in Table 2-1 for the world's six largest scientific and technological powers. When we look at the figures for all R&D expenditures, including those for defense and space, we find that there is a strong consistency in expenditure rates. Most the countries in the table devote about 2 percent of their GNP to R&D expenditures. The only country that deviates significantly from the norm is the USSR, which devotes 3.5 percent of its GNP to R&D. (Strictly speaking, the Soviet Union does not have a GNP, but rather a Gross Socialist Product. GNP is defined as the market value of goods and services produced in a country, but in a communist state, such as the USSR, market value is determined by the state, not by the market place.)

When we disaggregate the R&D expenditures somewhat and remove defense and space expenditures from consideration, we uncover an interesting distinction in the R&D spending behavior of the six countries. As table 2-1 shows, the percent of GNP devoted to R&D drops substantially in the cases of the United States, United Kingdom, France, and the Soviet Union, all of whom have substantial defense and/or space programs. However, the share remains virtually unchanged for Japan and West Germany, the two central protagonists of World War II, both of whom have eschewed a heavy defense commitment in the post-war era. More than a few commentators have suggested that the great economic strength of Japan and West Germany can be attributed, in part, to their lack of burdensome military outlays.

In the next to the last column of table 2-1 we have our indicator of scientific output: the number of papers appearing in some 2,400 of the world's most significant scientific journals. While the manpower and expenditure data reported here cover both science and technology, the publication indicator focuses soley on science. These publication data show that the United States is by far the world's leading scientific power, producing four times as many scientific papers as the second-place British. U.S. scientific papers, in point of fact, account for nearly 40 percent of the world's total output appearing in major international journals.

The Soviet Union and Japan produce far fewer papers than one would expect, given the size of their manpower pools and their R&D expenditures. In part, their low performance is accounted for by the fact that the world's central scientific literature contains few journals published in Russian or Japanese. However, it should be noted that English is the *lingua franca* of contemporary science. Japanese scientists are trained in English, and many Japanese scientific journals are actually published in the English language. Soviet scientists may not be as adept in English as the Japanese. However, another indicator of scientific output (not provided here)—the number of domestic journals published by a country—shows that the Soviets publish far fewer journals than the Americans. The general opinion of the scientific capabilities of the Soviet Union and Japan confirms the counts of scientific papers given in table 2-1: while both these countries have solid scientific capabilities, these are not exceptionally great.

The R&D Mix: Government Supported R&D

Government plays a substantial role in supporting R&D activities in all the developed countries, typically contributing from 40-60 percent of a country's gross expenditures or R&D. Once again we find West Germany and Japan set apart from the other countries. In West Germany and Japan, government support proportionately less research than in the United States, United Kingdom, and France. The smaller role of government in West Germany and Japan stems from the absence of substantial defense R&D outlays in these countries, for defense R&D is almost entirely government supported. In regard to the Soviet Union, it should be noted that because of the nature of a communist society, all R&D is, by definition, government supported.

The specific orientation of government-supported R&D for the five Western countries we are focusing on here is shown in table 2-2. This table portrays fairly substantial differences in orientation from country to country. While the orientation of each country has unique traits, the distribution of U.S. government funds for R&D appears to be the most unusual. The United States devotes a greater portion of its R&D budget to defense and space than any of the other countries in the table. Altogether, some two-thirds of the government budget in the United States is dedicated to these two categories. The U.S. government also devotes a greater proportion of its R&D funds to health research than any other country. While the United States expends a far greater share of its R&D budget on defense, space, and health, it spends proportionately far less than the other four countries in two R&D areas: economic development and the advancement of knowledge. R&D oriented toward economic development includes R&D directed toward developing improved capabilities in such areas as agriculture, construction, mining, transportation, and telecommunications. The U.S. statistic for advancement of knowledge given in table 2-2 is probably understated. A substantial quantity of basic research—research for the advancement of knowlege—is supported by the United States in defense,

Table 2-2
International Comparison of the Distribution of Government R&D Funds, Late 1970s
(*Percent of each government's total R&D*)

	Defense	Space	Energy	Economic Development	Health	Community Services	Advancement of Knowledge
France	33	5	7	21	4	2	24
Japan	2	6	10	22	3	3	54
United Kingdom	52	2	8	13	2	2	21
United States	49	12	13	9	11	4	4
West Germany	12	4	14	15	4	7	45

Source: National Science Board, *Science Indicators 1980* (Washington, D.C.: U.S. Government Printing Office, 1981), pp. 214-215.

space, energy, and health, and it is likely that the depressed figure for advancement of knowledge is a reflection of the fact that so much basic research is subsumed under the other categories.

Table 2-2 also shows what has already been noted: West Germany and Japan's R&D expenditures for national defense are very meager.

The R&D Mix: Industry-Supported R&D

The amount of R&D expenditures made by industry, taken as a percentage of the domestic product of industry (DPI), is provided in table 2-3. U.S. industry devotes the largest share of DPI to R&D, while Japanese industry devotes the least. The data suggest that American industry supports almost twice as much R&D per dollar of industrial product as the Japanese.

The source of funds for industrial R&D varies substantially from country to country, as can be seen in table 2-4. While nearly two-thirds of industrial R&D expenditures are generated by industry itself in the United States, United Kingdom, and France, virtually all Japanese industrial R&D funds and four-fifths of West German funds are industry-generated. Once again, the source of discrepancy is the substantial defense and space expenditures in the former countries. As these figures show, much of the government's military and space R&D in the United States, United Kingdom, and France is carried out by industry.

The Science Mix in the Developed Countries

Until quite recently, it was hypothesized by a prominent science historian that the kinds of scientific disciplines pursued within a country are similar for all those who have achieved a threshold level of scientific sophistication.[3] Thus, each of these countries would devote a similar proportion of its research

Table 2-3
International Comparison of Industrial R&D, Late 1970s
(*percent of the domestic product of industry*)

France	1.35
Japan	1.29
United Kingdom	1.75
United States	1.91
West Germany	1.64

Source: National Science Board, *Science Indicators 1980* (Washington, D.C.: U.S. Government Printing Office, 1981), p. 219.

Table 2-4
Industrial R&D by Source of Funds, Late 1970s
(percent)

	Industry	Government	Private Nonprofit	Foreign
France	66.5	25.3	0.3	7.9
Japan	97.9	1.9	0.1	0.1
United Kingdom	62.8	30.9	—	6.3
United States	64.7	35.3	—	—
West Germany	80.4	15.8	0.1	3.6

Source: National Science Board, *Science Indicators 1980* (Washington, D.C.: U.S. Government Printing Office, 1981), p. 218.

efforts to physics, chemistry, biology, and so on. The rationale for this hypothesis is that the different fields of science are interdependent. Advances in physics, for example, will lead to coattail advances in chemistry, or advances in biomedicine will not occur until certain bottlenecks in chemistry are overcome.

A detailed investigation of the scientific orientation of some sixty-two countries has shown that there is, in fact, a substantial variation in how countries distribute their scientific efforts.[4] However, these efforts are not distributed randomly, but rather fall into one of five basic patterns. The distinguishing characteristic of each pattern is a greater involvement in a specific kind of research than the world norm.

Pattern I contains a total of thirty countries. Their distinguishing feature is that they place a disproportionately heavy emphasis on biomedical research. The United States and the United Kingdom are pattern I countries.

Pattern II is comprised of fifteen countries. These countries focus primarily on research in the physical sciences. The Soviet Union and all Eastern European countries fall into this category. Japan is also a pattern II country.

Pattern III is composed of four countries. Pattern III countries distribute their research efforts between the biological and physical sciences more evenly than do pattern I and pattern II countries. Included here are France and West Germany.

Pattern IV countries, of which there are four, emphasize research on botany and the agricultural and food sciences. All the Pattern IV countries are less developed.

Finally, five countries are classified into pattern V. Pattern V countries focus heavily on both biomedical research and agricultural research. What is interesting about pattern V countries is that four out of five of them are former British colonies, including Canada and New Zealand.

Only three of the sixty-two countries do not fit into one of these five patterns.

Thus, the scientific mix in different countries varies. For the major scientific powers, we find that the United States and United Kingdom disproportionately emphasize research in biomedicine, the Soviet Union and Japan focus on the physical sciences (which includes physics and chemistry), and France and West Germany occupy a position between the biomedical and physical science orientations. It should be noted that because a country fits into a particular pattern, this does not mean that it neglects research typical of other patterns. For example, while the United States, as a pattern I country, devotes a disproportionate share of its scientific research effort to biomedicine, it is also the world's premier scientific power in physics.

Are American R&D Capabilities Eroding?

The statistics provided in the previous sections show the United States to be a giant scientific and technological power. Its budget for research and development is larger than the GNP of many countries. Its universities are acknowledged to be the best in the world. It is the only country capable of landing a man on the moon. Its scientists routinely win the great international prizes offered for scientific merit. Yet for all its obvious strengths, an uneasy feeling arose among Americans in the 1970s that national scientific and technological capabilities had peaked and that the country had entered into a period of decline. This feeling became so pronounced that toward the end of the decade President Jimmy Carter initiated a massive and expensive Domestic Policy Review to examine the causes and consequences of the decline of U.S. scientific, technological, and innovative capabilities.

How did it come to pass that the essentially optimistic American spirit was enveloped in this shadow of doubt? What led to the prophecies of gloom and doom? To what extent were the assessments of a decline in U.S. capabilities reasonable? To answer these questions, we must briefly consider a history of American scientific and technological achievements in the post-World War II era.

At the end of the war, U.S. might in science and technology was unquestioned and unparalleled. America dominated world science and technology like no other country in history. The horrible unleashing of atomic energy at Hiroshima and Nagasaki symbolized America's post-war strength in a number of ways. Militarily, no other country could summon the awesome power the Americans had encapsulated in their atomic bomb. Politically, the Japanese explosion elicited an unconditional surrender by the Japanese, bringing World War II to a sudden end. Technologically, the atomic bomb was the end-product of the costliest, largest, and most complicated R&D project in history—it was truly the product of a technological giant.

There are a number of explanations for America's total dominance of S&T at this time. In part, the U.S. postition was a consequence of the weakening by the war of the traditional S&T powers: England, France, and Germany. In 1945 the European economies were in shambles. Their first order of business was economic survival. In such an atmosphere, only enough S&T was supported as was necessary to keep the Western societies functioning.

The United States actually benefitted directly from the pre-war and wartime turmoil in Europe. After the ascendance of Hitler to power in Germany in the early 1930s, many of Europe's most talented scientists emigrated to the United States in order to escape the ensuing chaos. At the end of the war, the United States obtained further European talent when, along with the Russians, U.S. armed forces brought German scientists and engineers to the United States. Germans with a broad range of talent were co-opted into American service at this time; the best known were those scientists led by Werner von Braun, who formed the nucleus of America's new rocket program (which ultimately evolved into the space program).

While the weakening of Europe resulted in a corresponding strengthening of the United States, it should be noted that developments were taking place inside the United States that portended great things. The United States had a long history of fascination with science and technology. At its very inception, the nation was guided by men—such as George Washington, Benjamin Franklin, and Thomas Jefferson—who had a strong scientific and technical bent. As the United States developed economically, it faced a severe shortage of manpower, which it compensated for with indigenously developed machine power.[5] True to their democratic principles, Americans promoted universal education and were a highly educated people. Colleges, technical schools, and universities proliferated throughout the land. Americans who pursued advanced studies abroad returned home and dispensed their newly acquired knowledge and appreciation for good research practices to their fellow Americans. By the time of the outbreak of World War II, America had its coterie of home-grown scientists and engineers who ranked with the best in the world.

During the war years, a close alliance developed between American science and the military. This alliance was formed in order to further America's war aims. It led to the establishment of a pattern of mutual dependence between S&T and government that is still alive today. This alliance enabled the scientific and technological establishment to tap into a rich motherload of R&D funds. With access to a seemingly limitless supply of R&D dollars, S&T became big business in America.

In short, the United States was well positioned in 1945, partly because of the default of the traditionally strong S&T countries, and partially because of its own very real, substantial S&T capabilities.

American ascendancy in S&T continued for a quarter of a century. This is reflected in the fact that, during this time, so many Americans won the Nobel prize that it almost seemed an American prize. Then in the 1970s, cracks appeared in the country's S&T edifice. These cracks did not threaten to bring the whole structure down. But they suggested that the American scientific and technological system might be in decline. The most disturbing signs of decline included:

1. A steady decline in the percentage of GNP devoted to R&D. In 1965, U.S. R&D expenditure comprised 2.9 percent of GNP. By the end of the 1970s, this figure had steadily declined to 2.2 percent. In this same time interval, Japan, West Germany, and the USSR steadily increased R&D's share of GNP.

2. A decline in industrial R&D. In 1967, industrial R&D in the United States constituted 2.49 percent of the domestic product of industry. By the end of the 1970s, this figure was reduced to about 1.90 percent. Japan and West Germany increased their percentages between 1967 and the end of the 1970s.

The decline in U.S. industrial R&D has been attributed to a number of factors. The principal explanation is that U.S. corporations became increasingly oriented toward bottom-line results in this period. This financial orientation could be seen in the fact that the surest way to the top of many major companies was through the finance department. U.S. companies often tried to increase profitability through financial manipulations rather than through innovation. Conglomerates were formed at an increasingly rapid pace. With the onset of inflation in the early 1970s, R&D was trimmed because financial analyses showed that the present value of long-term investments—such as those associated with R&D—was low. Even traditionally strong R&D oriented companies, such as General Electric, succumbed to this financial approach to managing corporate growth.

3. A decline in government-supported R&D. Government support of R&D climbed inexorably upward in post-World War II America. Then suddenly in the late 1960s, as the Vietnam War and the massive social programs began to take their toll on the federal budget, government support of R&D was trimmed. Throughout the 1970s, most increases in government R&D spending were more than offset by the effects of inflation.

4. A decline in American patents and an increase in foreign patents. In 1966, 54,626 patents were granted to Americans by the U.S. Patent and Trademark Office. Toward the end of the 1970s, only 41,000 patents were granted to Americans each year. These statistics have been taken by some to represent a reduction in U.S. innovativeness, since patents are often viewed as an indicator of inventive activity.

Not only were U.S. patents that were granted to Americans declining in absolute numbers, but the number of U.S. patents granted to foreigners increased substantially, from 13,772 patents in 1966 to some 25,000 patents per year in the late 1970s. The share of U.S. patents going to foreigners increased from 20 percent in 1966 to some 38 percent in the late 1970s.

5. A decline in the U.S. share of the world scientific literature. The share of the world's scientific literature authored by Americans showed signs of decline in the 1970s. For example, in 1965, some 36 percent of the world's physics literature was authored by Americans, while by the end of the 1970s this share had decreased to 29 percent.[6]

6. A deterioration of American productivity. In the fifteen years from the mid-1960s to the late-1970s, U.S. increases in productivity (that is, output per man hour of input) were less than for any other major Western country, except the U.K. Japan, in particular, witnessed explosive productivity growth at this time and productivity increases in France and West Germany were substantial. There are many causes for productivity increases and declines. One is the extent to which new labor-saving technology is introduced into the economy.

7. A decline in American competitiveness. In the 1950s and 1960s, American technological products were recognized worldwide as being desirable and of high quality. Gradually, however, U.S. competitors— particularly Japan and West Germany—made inroads into areas where Americans had long dominated (for example, consumer electronics). By the 1970s, two of America's largest industries—automobile and steel —were in very serious trouble owing to their lack of competitiveness. As in the case of productivity, there are a number of factors that determine a country's competitiveness, and R&D is one of these. It is impossible to say with precision how much technology contributes to competitiveness, but it is clear that the prices of U.S. products could be made more competitive by the application of technology for cost-reducing purposes, for example, by using more industrial robots on assembly lines.

A superficial reading of these signs of decline in American S&T does seem to suggest trouble in Eden. However, a more careful examination of these signs shows that the message they convey is not really all that clear.

First, it should be noted that some of the apparent decline in U.S. abilities is relative and not absolute. The U.S. position in respect to Western Europe and Japan is certainly not as strong now as it was a couple of decades ago. This is natural and should not be taken as an indication that U.S. capabilities are deteriorating. America's total dominance of international S&T in the years following World War II was largely a result of Europe's destruction during the war. Europe had a three-century-old tradition as the world's primary producer of S&T. Europe was, after all, the originator of modern science. As to Japan, since the Meiji restoration in

1868, this country had pursued a policy of developing technological capabilities with an almost fanatical single-mindedness.

The United States could not hope to perform 40-50 percent of the world's R&D indefinitely. Once Europe had recovered from the scars of the war, it would naturally work to regain some of its lost capabilities. Similarly, dynamic Japan, with its tradition of discipline and perserverance, could not be expected to sit back and eschew S&T indefinitely. It was inevitable and natural that the other countries of the world would begin to catch up to the United States. No doubt, the gap separating the United States from the rest of the world will continue to close, even as U.S. R&D capabilities actually grow. However, as the statistics on Western R&D capabilities made clear earlier in this chapter, the United States remains far and away the single largest S&T performer. It must be recognized that the gains made by other countries do not *ipso facto* mean a deterioration in U.S. abilities.

Second, a number of the disturbing downward trends of the 1970s were reversed in the 1980s. For example, government support of research began climbing in real terms at the very end of the 1970s. Lackluster industrial performance in the 1970s caused many companies to reassess their finanacial orientation to growth, so that by the beginning of the eighties many of them had renewed their previous commitments to R&D, leading to industrywide increases in the proportion of sales dollars devoted to R&D. Indicators of U.S. productivity also improved somewhat at this time.

Third, some of the indicators of decline are misleading, patent statistics especially. Industry has been very dissatisfied with the U.S. patent system in recent years. For one thing, there is serious question of how much protection patents offer inventors, since litigation is so expensive that an inventor may not be able to protect his invention against infringement. For another, the process of obtaining a patent can be so time consuming that an invention's useful life can be over before the patent is finally awarded. Alternatives to patenting—such as maintaining trade secrets—have become increasingly appealing. So when the statistics show a decline in U.S. patenting activity, we must wonder how much of the decline reflects a real diminution of inventive activity, and how much is a result of inventors not bothering to patent their works.

As to the increase in the number of foreign patents in the U.S. patent system, this is most likely a consequence of the recent foreign penetration of U.S. markets, rather than an indication that foreigners have suddenly become amazingly fertile in their inventiveness. As foreigners introduce more and more of their technology into the American marketplace, they naturally want to protect their intellectual property by means of patents.

Notes

1. K. Polanyi, *The Great Transformation* (Boston: Beacon Press, 1944).

2. J.D. Frame, "National Economic Resources and the Production of Research in Lesser Developed Countries," *Social Studies of Science* 9 (1979): 233-246.

3. D.J. de Solla Price, "Nations Can Publish or Perish," *International Science and Technology*, No. 70 (1967): 84ff.

4. J.D. Frame, F. Narin, and M.P. Carpenter, "The Distribution of World Science," *Social Studies of Science* 7 (1977): 501-516.

5. H.J. Habakkuk, "The Economics of Labor Scarcity," in S.B. Saul ed. *Technological Change* (London: Methuen and Co., 1972).

3 Science and Technology in LDCs

The Colonial Experience

The fifteenth and sixteenth centuries comprised the great ages of exploration. For the first time in history men ventured forth from their homes in a conscious effort to explore and dominate the whole globe. Unlike the Hellens, Romans, and Mongols, who focused their attention on dominating the known world, these explorers sought the unknown. The first great explorers were the Portuguese and Spanish. Their motivations for conquest were God, glory, and gold, though not necessarily in that order. They hoped to convert the heathen to Christianity and bring their souls to God. They also hoped to bring glory to the monarchs who sponsored the explorations. And finally, they hoped to enrich both their monarchs and themselves by exploiting the conquered areas. Their explorations would eventually enable them to claim the whole of South and Central America (with some small exceptions in the Caribbean), substantial portions of Africa, and would take them to Asia and the Far East, where they had dealings with China and Japan.

Other European explorers followed the Portuguese and Spanish initiatives. The British, French, and Dutch founded substantial empires that took them to the four corners of the world. Initially, the proudest possessions of the British and the French were their territories in North America. The Dutch placed their major effort in Indonesia.

In the nineteenth century, there was a scramble to colonize the places on earth that remained unclaimed. Much of this activity occurred in Africa, where the Germans and Belgians joined the ranks of colonists (in Tanganyika and the Congo respectively).

The colonial experience was richly varied and complex. Some of the colonized areas were sparsely populated (for example, North America), others densely populated (for example, India). Some contained rather primitive societies (for example, Australia), others highly complex and ancient civilizations (for example, India). Some areas were disease infested and hostile to human habitation (for example, parts of tropical west Africa), others had climates similar to Europe's (for example, the central east coast of North America). In some colonies the colonizers had a profound social, economic, and political impact (for example, in all of Latin America), while in others the colonizers had only a superficial impact (for example, Egypt).

Considering the vast differences in colonial experience, one must be wary of generalizations concerning it. Nonetheless, one very significant generalization helps explain the plight of many of the less-developed countries today: in all cases colonization entailed the exploitation of the colonies by the mother country. The forms and intensity of exploitation differed from case to case, but the fact of exploitation remained nevertheless. For example, the Spanish conquest of Mexico and Peru entailed the deliberate destruction of existing sophisticated civilizations, the enslavement of the indigenous populations, and the single-minded expropriation of riches—primarily gold and silver—from the conquered territories. In contrast, the British suzerainty over North America entailed more subtle exploitation. The colonial subjects—who, with the exception of slaves and Indians, were of British origin—were required by law to supply England with raw materials that would be turned into manufactured products and sold back to the colonies. This approach was called mercantilism and formed the basis of the overall colonial system in the nineteenth and twentieth centuries.

Mercantilism kept colonies in a state of dependence upon the mother country. As in the North American case, the principal function of the colonies was to supply the mother country with primary products and markets for finished products manufactured from these primary products. Thus Malaya provided England with rubber, India with cotton, the Pacific islands with copra, and Nigeria with mineral wealth. All provided markets for British products.

By focusing on the production of primary goods, the colonies did not obtain the skills and knowledge necessary for industrialization. In fact, the mother country saw to it that industrial skills would not be developed. A well-known example of this is the successful British attempt to undermine the indigenous Indian textile industry, which at the time of the British arrival was flourishing.

The demise of colonialism was largely a product of World War II combined with the effects of social forces pushing for liberation. The British were greatly weakened by the war effort. The independence movement in India gained great momentum and led to the liberation of India in 1947. In other parts of Asia—Burma, Malaya, and Singapore—ties with the colonies were additionally weakened by Japanese occupation of these lands.

The Dutch faced even more serious problems. Not only was their great colony Indonesia occupied by the Japanese, but Holland was itself occupied by the Germans. Similarly, the French ties in Indochina were weakened at this time as a consequence of national liberation movements, Japanese incursions, and German occupation of France.

Within a few years of the conclusion of World War II, most vestiges of European rule had disappeared from Asia. In quick succession, the countries of North Africa gained independence, followed by a rush of decolonization in Black Africa in the early 1960s. By 1970, the only major colonial holdouts were the Portuguese colonies of Angola and Mozambique, and these would soon gain independence as a result of guerilla warfare activities. Portugal, among the very first countries to get involved in colonialism, was among the very last to abandon it.

With few exceptions, the governments of the decolonized countries were unprepared to cope with the problems of government in the modern world. Problems were particularly acute in Africa, where colonial territories were carved out at the convenience of the colonial powers without any rational basis. Thus, newly liberated countries might contain tribes that historically were implacably hostile to each other, so that any hopes for a smooth transition from colony to free state were bound to be shattered, as events in Nigeria, the Congo, and Uganda were to bear out so tragically.

While some countries were better prepared for liberation than others, in every case they were poor, technologically deficient, and burdened by high rates of illiteracy. The prospects for rapid economic development were not good.

It should be noted that not all the less-developed countries of the world are victims of European colonialism. Thailand, for example, never experienced European domination. Ethiopia, a country with one of the highest illiteracy rates in the world, had only a brief brush with European domination during the Italian invasion in the period between World War I and II. While Afghanistan and Iran have both experienced a British and Russian presence, neither one could be said to have been a colony of these countries. Other LDCs have been free from European domination for a long time. For example, all the major countries of Latin America attained independence in the nineteenth century. It is interestingly that a number of the world's most significant colonial powers in past centuries—Spain, Portugal, and Turkey—are themselves borderline underdeveloped countries today.

Clearly, not all the ills of the world can be blamed on the European colonial experience. Social, economic, and political factors that are independent of the colonial phenomenon can also make an important contribution to the developmental status of a country. It is certain that Ethiopia would be one of the least developed countries in the world today even without its brief exposure to Italian imperialism.

Steps toward Economic Development

Significant attention first turned toward the economic development of the poorer regions of the world in the 1940s and 1950s. At the outset, development was viewed as a relatively simple process. Countries were seen to be underdeveloped simply because they lacked capital. One way to look at things was to talk of vicious circles of poverty. Ragnar Nurkse proposed a vicious-circle scheme that looked like this:[1]

Proposition 1   Underdeveloped countries are poor.

Proposition 2   Because they are poor, they have a low propensity to save.

Proposition 3   Without savings, there is no capital.

Proposition 4   Without capital, there is no income-generating industry.

Proposition 5   Without industry, income remains low, which brings us back to proposition 1.

It was believed that one way to break out of this vicious circle was to provide capital to the less-developed country from the outside. With capital infusions, the country could begin to grow economically.

One reason for confidence in this approach is that it worked very well with the post-war Marshall Plan, which channeled capital from the United States to the war-torn countries of Europe. But it soon became apparent that this view of economic development was mechanistic and overly simplistic when applied to the less-developed countries. The European recipients of Marshall Plan assistance employed it effectively because they were already developed and could easily absorb the influx of new capital. The Marshall Plan helped put them back on their feet. It did not launch them on the path of economic development.

Capital began flowing into the LDCs. Frequently, factories would be built that could not perform the tasks they were designed to perform. Projects would be undertaken that would ultimately die on the vine for lack of adequately trained manpower. Factories would produce goods that had no markets.

As it became clear that the mere funneling of capital into underdeveloped countries was an inadequate development strategy, attention turned to the concept of technical assistance. The technical-assistance concept was based on the principal of helping the less-developed countries to help themselves. Thus, rather than simply pour huge quantities of capital into the LDCs, one should provide personnel in these countries with technical expertise and training as well. More and more attention was devoted to what was termed investment in human capital, as it became increasingly clear that the most significant shortage in less-developed countries was not capital but adequately skilled workers, technicians, and entrepreneurs. Investment in human capital boiled down to strengthening the education systems of the recipient countries and training personnel to obtain the skills necessary to function adequately in a technological world. Large numbers of promising (and not-so-promising) individuals were sent to developed countries en masse to receive advanced training in engineering, the hard sciences, economics, business, and the social sciences. In the United States alone, the Agency for International Development's (AID) Office of International Training financed the training of thousands of LDC participants in the United States, where they could make use of this country's excellent educational and training facilities. These participants would spend anywhere from several weeks to six years undergoing a wide variety of training experiences. Short-term training might involve on-the-job training at a milk processing facility. Long-term training would generally involve obtaining an academic degree at the bachelor's, master's, or doctoral level.

A number of problems have emerged as a consequence of this effort to upgrade human capital. These problems are not universal, but they are common enough to warrant some concern. First, and most significant, is what is formally called the migration of talent, but which is better known by the label *brain drain*. Students who have received advanced training prefer to live and work in developed countries, where working conditions are conducive to good work and the standards of living are high. In these cases, after receiving four years of advanced training, LDC trainees decide to remain in a developed country rather than return home.

Second, by receiving education in advanced countries, LDC personnel are often trained to handle advanced-country problems, which may or may not be relevant to conditions in the home country.

Third, advanced education can lead to the underemployment of highly trained individuals in their home countries, since there is often insufficient capacity to absorb the highly trained individuals into the domestic economic system.

Finally, investment in human capital is very costly and time-consuming. It takes twenty to thirty years to train a scientist or engineer, and during this time the trainee is consuming resources without making a notable contribution to the economy.

In more recent times, disappointment with the results of earlier approaches to economic development in the Third World have made development specialists recognize that the development process is far more complicated than was once conceived. One of the first individuals to recognize and articulate this complexity was the Swede, Gunnar Myrdal, who received a Nobel Prize in economics in 1974. In his classic 1968 study, *Asian Drama,* G. Myrdal points out that development occurs when there are the following six factors:[2]

Factor 1: Outputs and incomes must increase. At present we find most LDCs plagued with low labor productivity and low incomes.

Factor 2: Conditions of production must improve. Conditions in most LDCs are rather primitive and an adequate infrastructure for manufacturing is lacking.

Factor 3: Levels of living must improve. So long as they are depressed, economic development will be hampered. Currently, the levels of living of most LDCs are very low. Malnutrition, bad housing, and poor hygiene are pervasive. Educational and cultural facilities are generally weak.

Factor 4: Attitudes toward life and work must change. For example, modern societies require a certain level of worker discipline. Yet, in most LDCs worker discipline is low, as is exemplified by lack of regard for punctuality.

Factor 5: Institutions must change. One reason that the developed countries function smoothly is that their basic institutions are closely attuned to social and economic needs. LDCs, in contrast, often have flawed institutional systems. For example, many of them are saddled with antiquated land-tenure systems. They lack adequate credit facilities. Their public administration systems are weak.

Factor 6: Policies must improve. In particular, Myrdal points out that LDCs need to develop adequate planning capabilities.

In Myrdal's scheme of things, economic factors (1-3), noneconomic factors (4 and 5), and mixed factors (6) all play important and simultaneous roles in the economic development of a country.

The State of Science and Technology in LDCs

Scientific and technical personnel in less-developed countries produce only a tiny fraction of the world's R&D. We can appreciate this when we examine the publication of scientific papers in the world's 2,400 most important journals. Papers of LDC origin constitute roughly five percent of the world total. Over half these papers have Indian authors. When India is removed from consideration, LDCs account for slightly more than two percent of the total. The small scale of LDC research is further highlighted when we examine the average number of scientific papers produced per country. In Black Africa, this figure is 30 papers per country per year; in the Middle East (less Israel) it is 82 papers per country per year; in Latin America it is 119 papers per country per year; and in Asia it is 431 papers per country per year (52 papers when India is excluded). For purposes of comparison, consider the fact that the average Western developed country produces some 9,534 papers per country per year, while the average Eastern European country produces some 4,470. The output of scientific papers coming from the average Black African country is roughly equivalent to the output of papers generated by a typical American small college that has an active research program!

Unfortunately, data on LDC R&D manpower and expenditures are highly unreliable. The collection of good statistics is a costly process, and in LDCs other items have higher priority. We will examine some of these statistics for thirteen Middle Eastern countries in table 3-1. The statistics provided here are probably as good as exist for any LDC region. Nonetheless, they are still of limited reliability. An examination of the Middle Eastern data should give us a better feel for the dimensions of the R&D effort in LCDs in general.

All the countries in table 3-1, except Israel, are Islamic. Israel differs from these countries in many additional respects, all of which are rooted in the fact that Israel is basically a European country, while the others are clearly non-European. The inclusion of Israel in the table enables us conveniently to compare the R&D performance of a small European-type country with that of non-European LDCs.

The first column in table 3-1 provides us with the number of significant international scientific papers produced by the different countries. It shows that the scientific performance of the Arab countries lags far behind Israeli performance. Second place Egypt also dwarfs the other Arab countries. In the hierarchy of LDCs ranked according to the number of papers published,

Table 3-1
Indicators of Scientific and Technological Activity in the Middle East, Mid-1970s

	Scientific Papers Published	Scientists & Engineers	S&T Higher Education Enrollments	S&T Higher Education Degrees Conferred	S&T Expenditures (millions of dollars)
Algeria	43	—	13,941	832	$ 10.8
Egypt	720	225,348	130,937	14,867	73.2
Iran	173	76,693	55,401	4,822	83.9
Iraq	53	17,310	27,163	3,400	22.0
Israel	2,703	36,000	14,060	2,510	105.3
Jordan	15	758	1,712	102	6.3
Kuwait	15	4,102	1,097	69	1.1
Lebanon	127	12,000	4,482	617	11.2
Libya	9	8,232	3,623	237	—
Morocco	18	233[a]	5,960	249	—
Saudi Arabia	16	5,760	3,812	187	—
Syria	2	8,713	8,372	1,504	—
Tunisia	23	2,932	3,454	172	7.6

Source: J.D. Frame, "Measuring Scientific Activity in Lesser Developed Countries," *Scientometrics,* 2 (1980), p. 135.

[a]Includes only those engaged in R&D.

Egypt stands below first place India, with its output of 6,000-7,000 papers per year, and is roughly tied for the second place position with Brazil and Argentina.[3]

The second column gives figures for R&D manpower. It shows that with less-developed countries there may be little correlation between the size of the scientific and engineering manpower pool and scientific performance (measured here as the output of scientific papers). While expert opinion would hold that Israel is unquestionably the scientific and technological powerhouse for the region, it ranks a distant third in manpower. Its reported manpower pool is not as large as we might expect in comparison to other countries. For example, while Israel publishes some 180 times the number of papers published by Kuwait, it has only nine times the number of scientists and engineers. Such discrepancies may be due in part to the unreliability of the data; but in addition, they certainly reflect the fact that scientists and engineers in Western countries are far more productive than their LDC counterparts, owing to better training, better working conditions, and a host of additional factors that will be discussed later.

Columns three and four supply data on higher education in science and technology. These data correspond fairly well to the manpower data. This is not surprising, since scientific and technological manpower is largely a product of the educational system.

Finally, the last column contains figures on R&D expenditures given in U.S. dollars. Israel, Iran, and Egypt dominate this column. The other measures show that they are also the strongest R&D powers in the mid-1970s for the region. The funding figures appear to suggest that R&D performance is closely tied to the availability of funds for R&D. To put these figures into a different perspective, consider that a single American corporation— General Motors—devoted over $1.5 billion to R&D in a single year in the mid-1970s, far more than the total figure for the entire Middle East![4]

Appropriate Technology

In the early 1970s, a great deal of interest focussed on a number of related terms: intermediate technology, light capital technology, progressive technology, and appropriate technology. While these terms have slightly differing implications, they basically deal with the same thing: the perception that much of the technology available to LDCs and used by them is not really suitable to local conditions. Because what is really at issue is the appropriateness of technology, the term *appropriate technology* is the most commonly used.

E.F. Schumacher is the individual most closely associated with the concept of appropriate technology. The publication of his book *Small Is Beautiful* brought the ideas of appropriate technology to the public at large.[5] Schumacher observed two characteristics of LDCs that should have a strong bearing on determining the type of technology they should employ.[6] First, he noted that LDCs frequently suffered very high levels of unemployment. A technology would be appropriate to LDC needs then if it was labor intensive and created many jobs. Capital intensive technologies that were labor-saving were inappropriate technologies.

Second, Schumacher noted that LDC populations often fled the countryside and filled the cities. The cities were completely unprepared to cope with this influx. Consequently, living conditions, sanitation, public transportation, employment opportunities, and so forth, were inadequate. To Schumacher, this flight to the cities suggested that a technology would be appropriate if it could attract people out of the cities and back to the countryside. This meant that the technology should be designed to employ local material found in the countryside. Because the technology should be designed to fill local needs rather than national needs, it would entail rather small-scale production.

Thus technology was appropriate if it created employment, was geared to small-scale production, and was decentralized.

Westerners became excited by the idea of appropriate technology. In part, they looked upon it favorably because of the existence of so many highly visible and obvious cases of misapplied technology throughout the Third World. Here was an idea that made good sense. Westerners, both alone and in conjunction with colleagues from the Third World, began designing clever technologies that were well suited to conditions in LDCs. Many of these were designed to be built from local materials, held together with string, sealing wax, and rubber bands. Unfortunately, much of the leadership of the LDCs was not impressed with what it saw in these technologies. To many of them, appropriate technology became synonymous with second-rate technology. Some talk could be heard about appropriate technology being a developed country ploy to keep the Third World in a state of dependence.

The rationale behind the concept of appropriate technology is clearly sound. After all, no responsible development planner would want to support inappropriate technologies. In practice, however, defining what is appropriate and what is inappropriate can be rather difficult, and this is the chief problem with the idea of appropriate technology. To see how definitional problems can arise, consider the following hypothetical example: Thirdworldia is a poor country with few natural energy resources. It has no hydrocarbon wealth, no powerful rivers that can generate hydroelectric power, and no underground hot springs that could be used to produce geothermal power. Poor forestry and land management practices have led to substantial deforestation, so that wood and charcoal cannot be used to produce energy. Furthermore, it is cloudy over Thirdworldia eight months a year, so solar power is not a viable alternative. Given these conditions, what is an appropriate energy source for Thirdworldia? It can be convincingly argued that nuclear power is appropriate to conditions in Thirdworldia, even though it violates all the generally accepted rules for defining appropriateness. That is, a nuclear power plant is capital intensive, large scale, uses substantial quantities of highly complex imported technologies, makes minimal use of local materials, et cetera. Nonetheless, given Thirdworldia's unfortunate circumstances in regard to energy resources, a nuclear plant may be the only viable major energy source for the country. (This, of course, would have to be determined by means of a detailed cost-benefit analysis.)

Probably the greatest benefit arising from the recent interest in appropriate technology is that it forces development planners to consider the effects of a technology on an LDC's economic and social system. No longer do countries routinely opt for the most advanced technologies available, without giving any thought to the myriad consequences of such an action. The appeal of the concept of appropriate technology has even taken hold in the developed countries, as people recognize that technology should not be employed as if it were completely independent from its surroundings.

Obstacles to the Rooting of Science in LDCs

One need not be an expert on the Third World to recognize that there are many obstacles to the establishment and maintenance of scientific and technological capabilities in LDCs. Without question, it is more difficult for science to take root in an LDC than technology. Technology can be bought, borrowed, stolen, or developed indigenously. A good engineer is an individual who can build solid bridges, or keep a factory running smoothly, or design improvements in existing products and processes. A good LDC engineer can receive his engineering education in Europe, return home to his country, and spend a lifetime productively assisting in the development effort without doing any research. True, he must occasionally read the engineering literature to keep up with current developments, but involvement in research is not really necessary.

Scientists, on the other hand, justify their existence largely through the research they perform. Science entails measurement, observation, speculation, hypothesis testing—in short, the development of new knowledge. The undertaking of science requires the existence of conditions that foster the acquisition of new knowledge. The lack of these conditions leads to scientific activity that is seriously hobbled. A number of prominent obstacles to the rooting of scientific capabilities in LDCs will now be reviewed.

Lack of Resources

The lack of national resources that can be channeled to research is an obvious and important impediment to the development of Third World scientific capabilities. When we talk about the resources a country can draw upon to promote science, we are essentially discussing its capacity to afford such science. Science can be an expensive undertaking that LDCs, with their constrained budgets, may have difficulty supporting. Among the more prominently unaffordable enterprises are: a strong primary and secondary education system; a university system that is capable of producing scientists at the bachelor, master, and doctoral levels; research centers to employ scientists; libraries that maintain collections of the most current scientific literature; funds to pay for the actual conduct of research; funds to enable scientists to acquire advanced training abroad; funds to allow indigenous scientists to travel to international meetings and to visit with foreign colleagues abroad so as to minimize the isolation of the domestic scientific community; funds to convene meetings at home and to attract foreign scientific visitors. Until a country can afford these science necessities, a lack of resources will continue to be a prominent and possibly fatal obstacle to the fostering of healthy LDC science.

Isolation of Scientists

The isolation of LDC scientists has both an international and domestic side to it. Internationally, LDC scientists stand at the periphery of developments in science. They do not have the funds to attend world scientific conferences or to visit scientists in other countries. Their libraries are too poor to subscribe to the leading international scientific journals, or else when they do subscribe to them they arrive months late because of poor surface-mail handling. Foreign scientists rarely visit them. They lose touch with the faculty of the foreign institutions that trained them. Psychologically, they feel like second-class citizens in the international scientific community.

The domestic side of scientific isolation is more complicated than the international. There appear to be two aspects to domestic scientific isolation. First, the scientists in LDCs are often isolated from the rest of society, including its political and business leaders. Many LDC scientists feel that the issues of science are neither understood nor appreciated by members of their societies. As a consequence of this lack of understanding and appreciation, they do not receive much support from society for their efforts. One frequently hears calls coming from frustrated LDC scientists for nationwide science education programs that would build a mass base of support for indigenous science. Second, scientists are isolated from each other. There is little feeling of community among them. For one thing, their numbers are so small that they can hardly be called a community. For another, the hierarchical bureaucratic science systems that have evolved in many LDCs help to generate hostility among scientists.

Institutional and Bureaucratic Obstacles

It is frequently observed that science in underdeveloped countries becomes bureaucratized as it develops. Rigid hierarchies come into being and seniority becomes a strong determinant of the success of a scientist. Bold and individualistic thinking is discouraged, since in the bureaucratic milieu thinking by committee is commonplace, and bold thinking may threaten the status quo. In this environment, young scientiest, who stand on the bottom rungs of the hierarchical ladder, become disillusioned and, as a consequence, may intellectually stagnate or else may emigrate to an advanced country where their talents may be more fully utilized and appreciated. Furthermore, technicians—who are in critically short supply in LDCs—are poorly paid and viewed with little respect, making work as a technician an unattractive career option.

There are a number of explanations of why science becomes bureaucratized in LDCs. Probably the most obvious one is that science in underdeveloped countries is often overwhelmingly a government undertaking. In India, for example, roughly 85 percent of science is supported by the central government.

While bureaucratization of science is no doubt a consequence of government activity, the science bureaucracy frequently finds itself at loggerheads with the government, rather than functioning smoothly with it. There are essentially two causes underlying the science-government antipathy. First of all, in many areas of the world, especially in Africa and Asia, work in the civil service is more prestigious and remunerative than work in the teaching or research professions. Consequently—as many scientists bitterly complain—the nation's most talented young people may eschew science for the civil service.

Second, there seems to be a rivalry between scientists and government civil servants. The rivalry is seen primarily by the scientists, who feel underpaid, unappreciated, generally ignored, and disrespectfully treated by the government bureaucrats. One South Asian scientist expresses this outlook when he states: ". . . to keep someone with a string of university degrees at the end of his name waiting in the anteroom for an audience is manna to the hungry soul of the bureaucrat."[7]

The institutional and bureaucratic obstacles to the development of indigenous science in LDCs should not be downplayed. Insofar as these obstacles make for low morale and poor working conditions in research, talented young men and women will either not be drawn to careers in the sciences, or else, if they are, they will want to move from science to a more prestigious career in the civil service, or to leave the country entirely for more fertile research fields abroad. While these obstacles should not be downplayed, it should be noted that they are more closely associated with the maintenance of a strong indigenous science system than with the initial establishment of such a system.

Lack of Domestic Demand for Science

In most underdeveloped countries, the link between science and the economy is tenuous or nonexistent. Unlike the situation in the West, science has not proceeded hand-in-glove with economic development for several generations. Science is often imposed exogenously on LDCs. Like a mail-order suit jacket, it does not often set well on the shoulders of non-Western societies. One of the most crucial outcomes of this state of affairs is that in LDCs there is generally little domestic demand for science.

This lack of domestic demand for science has a number of important implications for the size and course of research in LDCs. For one thing, if there is little demand for science on the part of potential science users (for example, commercial, government, public health, and agricultural users), then it is unlikely that a scientific establishment will take root effectively in the developing society.

For another, without a domestic demand for science, it is difficult to direct research towards local and national needs. Consequently, LDC scientists frequently focus their research activities on problems of interest to the international scientific community, which usually have little bearing on LDC needs.

The Vested Interest of Elites

Occasionally, a major obstacle to the development of solid scientific capabilities in LDCs is the ruling group in a country. Attention has focused on this particular obstacle in recent years largely as a consequence of some egregious human rights violations perpetrated against intellectuals in Central America and the Latin American countries of Argentina, Brazil, and Chile.

One of the best articulators of the problem with elites is Amilcar Herrera of Argentina. In his writing, he does not deny that conventional views on the obstacles to the growth of science are largely true.[8] He identifies three conventional categories of obstacles to the domestic establishment and maintenance of science in LDCs: cultural obstacles, obstacles connected with the production system, and institutional obstacles. However, while admitting that these obstacles exist, he maintains that they are passive and not as crucial as other, more active, obstacles. This second group of active obstacles arises from the structure of society in LDCs, particularly in Latin America. At the very summit of Latin American society sit the old elites who jealously guard their social, political, and economic prerogatives. They are unaware of what science can offer and fear it because they believe it will disrupt the status quo.

In recent decades, some of the elites' power has been eroded by a newly emerging middle class. Herrera maintains that although one might suspect that the members of this middle class would hold liberal views and encourage the growth of indigenous science, they—like the old elites—do not understand science and resist its establishment for fear that it will disturb the status quo, in which they have a great deal at stake. Thus, the middle class has allied itself with the old elites in its resistance to indigenous science. Since these two classes control the political processes in Latin America, governments there do very little to promote the rooting of R&D within their borders.

Herrera and other writers point out that science in Latin America is strongly affected by political and ideological considerations. If a scientist does not subscribe to the views of the ruling regime, he may very well find himself unemployed—or, as we have seen in Argentina in recent years, he may be jailed or may simply disappear.

Problems with the politicization of science exist in many LDCs outside Latin America as well. The prevalence of this phenomenon is to a certain extent the consequence of a lack of appreciation among denizens of the Third World of the functioning of science in society. Good science simply cannot be legislated in a parliament or be dictated by an autocrat. For good science to be performed, researchers must have the freedom to pursue whatever avenues of investigation are required to solve a problem. This fact is recognized even in the Soviet Union, a highly ideological, politicized society. Scientists in the physical sciences have as much autonomy as any group of individuals in the USSR. As a consequence, the Soviet Union is able to produce some first-class physical science research. In contrast, Soviet life sciences and social sciences—which are not free from ideological interference—are generally acknowledged to be third-rate.

Cultural and Cognitive Obstacles

It has often been observed that people in the Third World frequently lack a scientific outlook, and that this is a serious impediment to the establishment of solid scientific capabilities in LDCs. It seems that every Western visitor to the Third World has his favorite anecdote that illustrates this point—about the Ph.D. scientist who consults his astrologer regarding the timing of a large investment; or about the managers of a milk processing plant who observe stringent sanitary regimes within the plant, only to have the milk distributed in filthy buckets once it leaves the plant; and so forth.

Certainly, the lack of a scientific outlook in LDCs can be remedied to a certain extent through education: Third World students can be taught the importance of observation, the scientific method, objectivity, and all the other accoutrements of the proper scientific outlook. However, there is evidence that different cultures actually perceive reality differently, and that these culturally rooted perceptual differences are stubbornly resistant to change through simple indoctrination. Inculcating a scientific outlook in LDC societies may be a very difficult task indeed.

Differences in Western and Third World approaches to reality were illustrated by Francis Dart and Panna Lal Pradhan in a well-known study that examined the receptivity of Nepalese children to a scientific education.[9] These two researchers interviewed Nepalese school children in depth and came up with a number of interesting findings. For example, they determined that the children, despite their Western-style education, still adhered to traditional ways of explaining the occurence of natural phenomena (for example, some children maintained that the earth sat on the back of an elephant, and earthquakes resulted when the elephant shifted his weight). However, alongside the traditional answer, the children would give a scientific response that was learned in the classroom (the children noted that pressures in the earth's molten core contributed to stresses in the crust that resulted in earthquakes). The mechanical way in which the scientific explanations were given led the researchers to believe that the children did not really follow the reasoning behind such explanations, but were merely regurgitating what they had learned by rote in the classroom.

Dart and Pradhan also found that the children saw the control of nature as occurring through religious ritual and by magical means.

Finally, the researchers also determined that in the world view of the Nepalese children, knowledge is immutable, something that exists in a fixed quantity and is handed down from generation to generation. The researchers explicitly inquired how knowledge hitherto unknown to anyone might be acquired or how it might be sought. The children responded that they had been taught that new knowledge was not to be expected. Dart and Pradhan concluded that, for the subjects of their study, the source of knowledge is authority, not observation.

These findings on the scientific outlook of Nepalese children stand in sharp contrast to those obtained in an examination of the outlook of a control group of American school children who explained natural phenomena by scientific means, who believed that men could control their environment by direct action, and who saw observation as the source of knowledge. Tests conducted by Dart and Pradhan also showed that the American school children displayed a higher capacity for abstract thought than their Nepalese counterparts.

Robert Solo expressed parallel feelings when he pointed out that people in LDCs frequently lack certain cognitions that are found in the West.[10] He identifies four cognitions: 1) the cognition of mechanism, which requires an understanding of the machine and all of its implications; 2) the cognition of technique (technical skills), which goes beyond the mere appreciation of the machine and applies this appreciation to making the machine work; 3) the cognition of process, which entails the ability to organize all the complexities of society into a viable process; and finally, 4) the cognition of science, which sits at the apex of the pyramid built by the three previous cognitions and whose chief focus is on the mastery of science.

Conclusion

For the most part, the scientific and technological capacities of the Third World are very weak and are likely to remain so in the forseeable future. In recent times, technology has been viewed as an engine of growth, and so long as LDCs remain technologically weak the prognosis for their development is not very good. There is a bright spot, however. The 1970s witnessed the sustained economic growth of a number of LDCs at a high level—the so-called newly industrialized countries (NICs). Not surprisingly, parallel with their economic advances, these countries are developing the germ of solid scientific and technological capabilities. They stand as an example to the poorer countries, showing them that it is indeed possible to raise oneself out of the quagmire of underdevelopment.

Recognizing the importance of technology for development, the LDCs have been making rather strident demands for better access to the technology of the developed countries. This is one of the cornerstones of the LDC call for the establishment of a New International Economic Order. It is important to understand these demands and the developed countries' responses to them, because they are a major source of friction between the rich countries of the North and the poor countries of the South. Also, if the demands are met, it will have a major impact on international business. The Third World demands for developed countries technology will be discussed in detail in chapter 10.

Notes

1. R. Nurkse, *Problems of Capital Formation in Underdeveloped Countries and Patterns of Trade and Development* (New York: Oxford University Press, 1967), p. 5.

2. G. Myrdal, *An Approach to the Asian Drama* (New York: Vintage Books, 1969), pp. 223-228.

3. J.D. Frame, F. Narin, and M.P. Carpenter, "The Distribution of World Science," *Social Studies of Science* 7 (1977):507.

4. "Annual R&D Spending Scoreboard for U.S. Industry (1978)," *Business Week,* 1979 4, July, p. 53.

5. E.F. Schumacher, *Small Is Beautiful: A Study of Economics as if People Mattered* (New York: Harper & Row, 1973).

6. E.F. Schumacher, "The Case for Intermediate Technology," in G. M. Meier, ed., *Leading Issues in Economic Development,* 2nd ed. (New York: Oxford University Press, 1970), pp. 355-359.

7. B.R. Seshachar, "Problems of Indian Science since Nehru," *Impact of Science on Society* 22 (1972):141.

8. A. Herrera, "Social Determinants of Science Policy in Latin America," *Journal of Development Studies* 9 (1972): 19ff.

9. F.E. Dart and P.L. Pradhan, "Cross-Cultural Teaching of Science," *Science* 155 (1967):649ff.

10. R. A. Solo, *Organizing Science for Technology Transfer in Economic Development* (Landsing, Mich: Michigan State University Press, 1975), pp. 37-40.

4 R&D and the Firm

Introduction

Technology does not just happen. Projects must be selected, funded, and staffed. Working ideas must be taken from the laboratory bench and scaled up for production. Industrial R&D efforts must be coordinated with the firm's other activities.

In this chapter, we examine several aspects of the environment in which technology is created within the firm. Such a review of the place of R&D in the firm elucidates how companies go about the business of research and development.

The Role of R&D in the Firm

Half the R&D performed in the United States is funded by industry. (The other half comes from government.) The question naturally arises: Why does a firm such as General Motors expend more than two billion dollars on R&D in a year? Why do thousands of other companies, in a variety of industries and of different sizes, spend scarce resources to support activities whose outcomes are highly uncertain?

Firms support R&D for a wide variety of reasons. Some see it basically as an investment. Resources are dedicated to activities today in the hope that they will yield major pay-offs tomorrow. Clearly, this is how the Ambac division of United Technologies views its R&D efforts on a new superefficient diesel electronic fuel-injection system for automobiles and light trucks. Although R&D work began on the system in the late 1970s, it is anticipated that the product will not come to market until the 1990s![1]

Other firms see R&D as a necessary business operating expense. In their view, R&D efforts must be supported as a matter of course for survival in our technological age. R&D can lead to innovations that reduce production costs or that enable the firm to introduce new products into the marketplace or that allow it to match competitors' new products. It is a necessary expense for business operations, just as legal and accounting costs are necessary expenses.

The U.S. Internal Revenue Service recognizes this dual nature of R&D and allows corporations to treat R&D either as a capital asset for tax purposes or as a business expense.[2] When treated as a capital asset, R&D costs are depreciated over a defined period of time, while as an expense they are fully deducted in the year they are incurred.

The specific role of R&D in a firm is determined by a number of factors. First, the nature of the industry in which the firm functions is an important consideration. Some industries are very fast paced—the semiconductor, computer, and telecommunications industries are examples. In such industries, the useful life of a new product is often measured in terms of one or two years, or occasionally, a few months. The industry may be so fast paced that it is hardly worth obtaining patent protection for new inventions, since the inventions will be obsolete before the patent filing procedure is finished. Other industries are rather mature and slow paced. The tire industry has traditionally been a rather slow-paced industry with major innovations being introduced every fifteen years or so. Clearly, in fast-paced industries a major commitment to R&D is necessary for survival, while in slow-paced, mature industries it is not. Having made this generalization, it should be noted that even so-called mature industries can be shaken to their foundations by the introduction of new technologies. For example, Wilkinson's introduction of the stainless steel shaving blade into the mature razor blade industry was a shock that set all the major producers of razor blades onto a course of action that led to an increased commitment to R&D and innovation.

Second, the corporate ethos will be important in defining the R&D role in the firm. If the corporation's top executives see their firm as an innovator and pathfinder, the commitment to R&D is likely to be high. If, on the other hand, they see the firm as a plodder whose motto is slow and steady wins the race, then R&D will likely play a minor role.

Third, environmental conditions help shape the role of R&D in the firm. For example, in inflationary times, firms tend to downgrade long-term R&D and focus on incremental innovations that give very short-term results. The reason for this is that financial analyses show that the present value of long-term investments is low in highly inflationary times. Government regulations are another environmental factor that may shape the role of R&D in the firm. The U.S. steel industry, for example, complains bitterly that clean air requirements cause it to direct its R&D efforts to reducing harmful emissions from the steel manufacturing process rather than to improving steel-making productivity. Other environmental factors that have an impact on what role R&D plays in the firm include its geography (for example, Is it near a university?), antitrust laws (for example, Will the laws forbid a joint venture?), and the availability of capital (for example, How will an equity based company raise capital for new ventures when the stock market is depressed?).

Fourth, corporate capabilities will help determine the role of R&D. Such things as the corporate cash position, the quality and quantity of manpower, and the quality of laboratory facilities will contribute to defining the direction of R&D in the firm.

The fifth and final factor that will be considered here is R&D strategy. R&D strategy both shapes the R&D role and is in turn shaped by it. A firm can pursue a number of different R&D strategies, either singly or in combination with each other. These strategies are not mutually exclusive and occasionally overlap. They include:

Offensive (First-to-Market) Strategy

An offensive strategy requires a high commitment to R&D. With an offensive strategy the firm's goal is to come up with a new product, and to arrive at the market place with this new product as the first with the most. By being first to market, a firm hopes to be entrenched in the marketplace by the time competitors enter into the picture, so that it can maintain a substantial market share. In addition, if the firm has produced an exciting new product, it may be able to pursue a cream-skimming pricing policy before the arrival of competitors. That is, it may be able to charge what the market can bear and quickly recoup much of its R&D expenses.

Of course, an offensive strategy is a high-risk strategy. To be first to market requires breaking new ground. It may well be that once the product is brought to market, its intended consumers reject it. Studies have shown that the single largest cause of failure of R&D efforts is the marketplace, and those companies that adopt offensive strategies are fully aware of this possiblity.

Defensive (Follow-the-Leader) Strategy

A firm that pursues a defensive strategy is not concerned with being first to market. As the name implies, a defensive strategy is basically a cautious strategy. The firm is willing to allow someone else to test the waters of the marketplace. Thus, it is willing to forego some market share in the early stages of the innovation's product life cycle. If it appears that the product is successful in the marketplace, then the firm directs its R&D effort to duplicating the innovator's product. It may add a new twist to the product in order to avoid patent-infringement problems as well as to make its product more attractive to prospective consumers. A follow-the-leader firm often has good production skills that will enable it to reduce production costs and make its product highly competitive. While an offensive strategy entails high risks, a defensive strategy is generally low risk.

Licensing

Licensing is a special case of defensive strategy. By licensing a technology, a firm is able to take advantage of another firm's R&D efforts. Consequently, it is able to reduce the expense of maintaining strong R&D capabilities in-house. Furthermore, the licensed technology has, in all likelihood, already shown itself to be marketable and production problems have been ironed out, so that the licensing firm has largely eliminated the considerable risks associated with new-product introduction. While a firm that follows a technology-licensing strategy need not have a high commitment to R&D, it must nonetheless have a certain level of technological competence in order to produce the licensed product and to maintain quality control. It is particularly important that the licensee have very good production capabilities or that it have a comparative advantage in cheap labor, since it will gain market share largely on the basis of low price. It has often been noted that this strategy characterized much of the Japanese industrialization effort in the decades following World War II.

Patch-up Strategy

With this approach a firm extends the life of a product by making cosmetic improvements to it. For example, a minicomputer firm may attempt to increase the appeal of its rapidly obsolescing computer by increasing the computer's memory capacity, rather than by redesigning the whole system. Since a clear objective of this strategy is to reduce R&D costs, patching-up reflects a relatively low commitment to R&D.

The patch-up approach is a short-term strategy. If it works, it extends the life of a product for a while. However, in the long run, this approach will not provide a firm with the significant innovations that may be needed to survive in a world of rapid change.

Niche-Filling Strategy

The principal objective of this strategy is to avoid head-on competition with firms that are entrenched in a market. A firm pursuing this strategy looks for unfilled niches in the product lines of other firms, and then produces goods to fill these niches. For example, a company wishing to enter the computer market in the 1960s and 1970s was faced with the unpleasant prospect of competing directly with the supergiant of the industry—IBM. If the company determined that the only way it could survive in the computer industry was to look for unfilled niches in the IBM product line, it could launch itself—as many firms did—into the production of minicomputers (which IBM ignored at the time, preferring to focus on mainframe computers) or computer peripherals (for example, printers) that were compatible with IBM products.

The niche-filling strategy is clearly appropriate to a small new company that is just beginning operations. Such a company has sufficient problems raising capital, organizing its operations, developing products, etc. without worrying about being wiped out by entrenched giants who can afford to underprice and out-market it. The strategy is appropriate to more established firms as well. The 3M company's basic R&D strategy is based on filling product gaps and avoiding direct confrontations in the marketplace with competitors.[3] In general, firms that pursue a niche-filling strategy are interested in coming up with innovations, so they generally have a high commitment to R&D.

Maverick Strategy

A maverick strategy is one where a firm develops an innovation outside its usual line of business. This strategy may entail a one-shot deal, where a firm recognizes that a new product it has developed—or can develop—will be successful in a market into which the firm has never before entered. Or else, the strategy may be pursued systematically and continuously. Du Pont, for example, has set up a new products department that examines corporate technological achievements carefully and attempts to find new markets for them. The department has been particularly successful in establishing an instrumentation division that markets various instruments that Du Pont has developed for internal purposes.

Acquisitions Strategy

A firm that is anxious to nuture R&D capabilities quickly can do this by acquiring organizations that already have developed these capabilities. One of the most dramatic examples of this strategy in recent years has been the wholesale acquisition by United Technologies of many firms that are technology intensive (for example, Otis Elevator, Sikorsky, Carrier, Mostek). The purpose of United Technologies' acquisitions policy has been to purchase companies that can benefit from the corporation's R&D capabilities and can, in turn, contribute to these capabilities as well. They hope that the whole of the different divisions' R&D skills is greater than the sum of the individual parts.[4]

Acquisitions strategies have been heavily criticized in recent years. They have been blamed for contributing to declining innovation in the United States. It is argued that acquisitions are fundamentally financial transactions that do not create new R&D capabilities, but rather detract from investment in R&D. By the outset of the 1980s, many corporations budgeted more funds for acquisitions than they did for R&D.

Acquisitions can be made not only of companies, but of individuals. Thus, a company that wants to strengthen its R&D capabilities may pursue a strategy of hiring talent away from other companies, universities, or government.

Organization of R&D in the Firm

There are four basic line functions that a typical firm must accommodate: production, finance, marketing, and R&D. (A fifth basic function—personnel—is here viewed as purely administrative and will not be discussed.) Production entails the physical act of producing something that the firm hopes to sell. At one time, firms generally produced tangible goods. Today, they produce more services than goods. In either case, they must be concerned with such things as scheduling tasks, carrying out projects, and assuring the quality of their products during the production phase.

Finance activities are directed at raising and managing funds. These funds are most frequently raised by offering equity interest in the firm (through the issuance of stock) or through debt.

Marketing involves a wide range of activities, including the conduct of market surveys, making sales projections, pricing products, and promoting products in the marketplace.

R&D entails the application of scientific and technological know-how to improving products and processes. When the R&D effort focuses on developing new products, the results of the effort are called *product innovations*. When R&D focuses on improving production processes, the results are called *process innovations*.

Clearly, the basic line functions listed here are not carried out in isolation from each other. In the effective organization, they do, in fact, heavily interact with each other. Production runs, for example, are calculated on the basis of sales estimates made by the marketing group, and sales promotion budgets must be made with one eye on financial constraints. The R&D group in particular needs to work very closely with the other groups. R&D cannot be permitted to be undertaken in a vacuum, or else the resulting discoveries are bound to be irrelevant to corporate needs. R&D must be aware of production requirements and discoveries must be tailored to the production capabilities of the firm. A new invention that can be readily built at the laboratory bench but cannot be mass produced is not likely to be very useful to the firm. R&D must be aware of financial requirements since the R&D project portfolio will largely be shaped by budget factors. Finally—and perhaps most important—R&D should work closely with the marketing department. In determining future products that the firm should produce, the marketing people must have a good idea of what the R&D staff has done in the past, is doing today, and is capable of doing in the future. On the other hand, in order to avoid inventing products that have no customers, R&D should heed the marketing department's advice on what will and will not sell.

Unfortunately, while everyone recognizes that close coordination of the different basic functions is an important condition for the smooth operation of the firm, in practice, coordination is usually less than ideal. This is a consequence, in part, of the suspicions of the different groups toward each other. R&D people are often viewed by the other groups as ivory-tower types with their heads in the clouds. R&D staff, on the other hand, tend to see the other groups as technically untutored, and overly concerned with making a buck and conforming to bureaucratic niceties.

One of the most fundamental tasks a firm must perform is to organize itself to carry out these basic functions in an efficient manner. The organization of the R&D function can be pursued in a number of different ways. Only the most common organizational forms will be discussed here. It should be noted that these organizational forms are not mutually exclusive, and in industry many hybrid versions exist that combine different features of each.

R&D Organized According to Scientific Discipline

One obvious way that R&D can be organized in a firm is according to scientific discipline. This organizational form is portrayed in figure 4-1 for a hypothetical engineering firm. In this firm, R&D is broken down into three engineering disciplines: chemical engineering, mechanical engineering, and electrical engineering. Engineers in each of these groups work on problems that bear on their narrow areas of expertise.

Most scientists and engineers would be quite comfortable with a disciplinary breakdown, since it coincides nicely with what they grew accustomed to during their years of undergraduate and graduate education. The disciplinary organization of R&D in the firm is simply a transfer to the firm of the university's organizational structure.

One advantage of this approach is that it enables engineers and scientists to keep well informed of developments in their specialities, since they constantly rub shoulders with colleagues in the group who have like in-

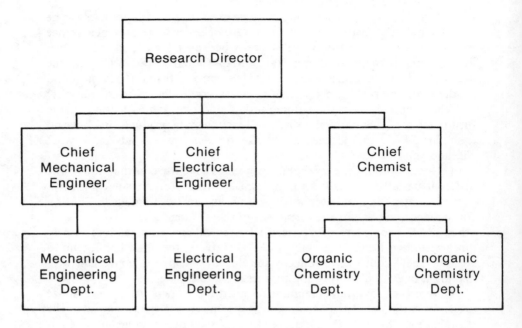

Figure 4-1. R&D Organized According to Scientific Discipline

terests and training. The individuals in the disciplinary group tend to read the same technical journals and attend the same technical conferences. In such an environment, the engineer or scientist continuously receives or disseminates information on the latest developments in his area of expertise. He is far less likely to become a victim of obsolescence in this environment than in one where he is isolated from like-minded colleagues.

This advantage of the disciplinary approach is also one of its principal weaknesses. Today we live in a world of specialists, where the individual specialist knows more and more about less and less. No individual has expertise in more than a handful of areas. Yet, complex problems do not recognize disciplinary boundaries. Their solutions often require expertise in a wide range of areas. An R&D effort divided along disciplinary lines may well lead to disciplinary inbreeding and a lack of the multidisciplinary cross-fertilization of ideas needed to come up with satisfactory solutions to complex problems. The history of S&T is repelete with examples of technical problems being resolved through interdisciplinary efforts. One of the best known of these is Crick (a biologist) and Watson's (a physicist) determination of the double helix structure of the DNA molecule.[5]

R&D Organized According to Product Division

Firms are often organized according to product lines, where distinct divisions are established within the corporation to deal exclusively with these product lines. The R&D effort can readily be structured to dovetail with this organizational form. Each division is given its own R&D group that is particularly familiar with divisional products, problems, and capabilities. Figure 4-2 shows two possible ways that R&D can be organized into divisions for a hypothetical electronics firm. In the top part of the figure, R&D is housed under a single roof, with specific groups within the centralized R&D structure responsible to individual divisions. In the bottom part of the figure, the operating R&D groups are actually housed within the separate divisions. Here overall R&D coordination is effected in a small office that contains the R&D director and his immediate staff.

The principal advantage of the product-line organizational form is that it enables the R&D effort to be responsive to divisional needs. Consequently, problems of reconciling the overall goals of the R&D staff with those of the divisions are minimized. One disadvantage of this organizational form is that in responding to the needs of the divisions, the R&D that is performed will primarily be for the short term, since most of the divisions' needs will be directed toward current or near-term problems. Another disadvantage is that R&D will take on a rather fragmented divisional outlook and corporatewide R&D opportunities may be missed.

Central Research Laboratory

An organization may choose to establish a central research laboratory (or laboratories) that addresses itself to long-term and corporate-wide problems. Major technology-based corporations typically have such laboratories. Perhaps best known among these is IBM's Watson Laboratory and ITT's Bell Laboratories. The central laboratory is often physically located apart from the rest of the organization, sometimes in futuristic facilities in a bucolic setting. The atmosphere of such a laboratory is redolent of a university campus.

This type of central laboratory is supported by firms that recognize that large gains are to be had by pursuing long-term research and research that is not specifically keyed to the needs of corporate operations today. Such research is enormously risky, since it explores unmapped terrain. By the same token, successfully pursued long-term research can be enormously profitable because it creates entirely new products and processes, giving the inventor complete control of the market during the early stages of the invention's introduction into the marketplace.

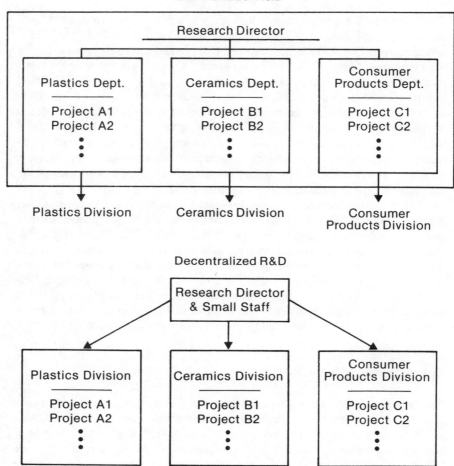

Figure 4-2. R&D Organized According to Product Division

A drawback to the central laboratory is that it is rather far removed from market forces. Its ultimate objective is to come up with new products and processes that will generate income for the firm. If the lab is too isolated from the marketplace, the firm may not be able to effectively translate new inventions into profitable innovations.

Another drawback is that the central lab, operating as it does in a rarefied intellectual atmosphere, may lose sight of the business it is supposed to serve. In pursuing knowledge for knowledge's sake, it may forget that its primary function is to help the firm survive and profit in a highly competitive world.

The possible shortcomings of a central laboratory are generally well recognized. A firm that supports such a lab must have faith in the efficacy of long-term research and a willingness to assume substantial risk, and it must have adequate resources to finance an operation with a rather long payback period.

Venture Management

It is generally acknowledged that major innovations occur most frequently in a fluid, freewheeling environment such as exists in relatively small technical companies. Large corporate environments, with their attendant bureaucracies, tend to stifle creativity and innovation. One piece of evidence that is often cited to support this view is that all but one of the major innovations in petroleum cracking came from fairly small, independent outfits. The innovations did not originate in the major oil companies, with their billions of dollars of assets.

Recognizing that bigness may be inimical to innovation, large firms such as Du Pont and 3M have established R&D systems that duplicate a small company environment.[6] This approach is called venture management. It enables entrepreneurial types within the organization to run projects with minimum interference from above. Venture managers are given a budget and freedom to hire staff. Criteria are established to assess (after a stipulated period of time) whether the venture is worth pursuing on a large scale. Once the ground rules have been established, the venture manager is on his own and has enormous freedom to choose his own course of action.

A firm that is committed to venture management is clearly opting for a highly decentralized R&D system. For the system to function effectively, corporate planners have to balance the need for corporate control over activities against the desire to create an atmosphere conducive to innovation.

Matrix Management

The matrix system of organization creates a labor pool of scientists, engineers, and technicians who can be tapped for service on different projects. Individual R&D personnel look to the head of their technical department for professional guidance. Technical departments see to it that their staff are well trained, fairly paid, perform their tasks properly, and so forth. If a project leader needs someone with certain technical skills to work on his project, he obtains the needed individual from the technical group that covers that skills area. When technical staff work on specific projects, they are responsible to the project leader for their performance on the project.

The major advantage of the matrix approach is that it uses technical personnel efficiently. For one thing, this approach recognizes that many technical specialties have only a limited role to play in a given project. For another, by separating technical workers from projects, stability is introduced into the overall personnel system. For example, when a project is terminated, it is not necessary to scramble to find a slot for displaced personnel on other projects, or worse yet, to fire them. Rather, personnel are hired and retained according to the overall need for them across many projects.

The most obvious disadvantage of the matrix approach is that it violates one of the fundamental precepts of good management, that is, that lines of authority be clearly delineated. From the point of view of the scientist working on a project, lines of authority are somewhat muddled. On the one hand, he clearly has to please his boss in his technical department, since this individual plays a strong role in determining his salary level, promotional possibilities, professional development, etc. On the other hand, his immediate boss on the project is the project leader, who must also be pleased. In practice, if the project leader, technical leader, and the scientist approach their work in a cooperative spirit, the likelihood of conflict arising from the dual lines of authority is minimized.

Another disadvantage of the matrix approach is that if the technical worker is constantly shifted from project to project, he may not identify closely with any of them. Without project identification, he is not likely to have a great commitment to the success of the projects he works on.

Determining the Level of Support for R&D within the Firm

An important task facing the firm that maintains an R&D program is to determine how much material support should be dedicated to it. The level of support will largely depend on the overall R&D intensity of the firm's industry, the firm's commitment to R&D, and its R&D strategy.

R&D intensity varies dramatically from industry to industry. The U.S. semiconductor industry, for example, dedicates over 7 percent of its sales to R&D, while the steel industry dedicates only 0.6 percent.[7] If a firm wants to remain competitive in its industry, it will set its R&D level near the industry norm.

The firm's commitment to R&D and its R&D strategy will help determine whether the R&D budget is above or below the norm. Some firms whose sales are steady owing to a few products with high consumer loyalty are content to support the bare minimum R&D effort. Their approach is to sit back and milk the cash cows. Other firms in the same industry may be run by management teams who are forever trying to expand market share and capture new markets. The strategy of these firms would likely lead them to display a high commitment to R&D.

Determining the industry norm for R&D expenditures is rather easy in the United States since the Security and Exchange commission (SEC) requires publicly held firms beyond a certain minimum size to file a 10-K report that provides summary data on their R&D expenditures. These figures are gathered and conveniently reported for some 800 major firms each July in an issue of *Business Week* magazine. The statistics have to be interpretted with some caution, however, since, despite SEC guidelines on what may and may not be included in the 10-K statement, there is substantial variation among the firms in how they define R&D.

There are a number of ways a firm can establish an overall R&D budget. Four different ways will be described here.

Support All Good R&D Projects

With this approach, an organization assesses prospective projects according to various criteria (for example, technical merit, marketability). All projects that are viewed to be meritorious are supported without regard to cost. The overall R&D budget will vary then with the cost of the supported projects, which will change each year. This approach can be adopted by a firm where money is no object or by a firm that supports so few projects that it can easily afford to bankroll the two or three meritorious ideas it comes across each year.

Base the R&D Budget on the Previous Year's Budget

Another way to establish an R&D budget level is to base this year's level on last year's, adjusting the budget upwards, perhaps, to accommodate inflation. Such an approach obviously introduces an element of stability into the budget, since it will not lead to dramatic year-to-year fluctuations. R&D planners will have a good idea of what budget levels will be two or three years down the line and can plan their R&D portfolios accordingly. However, if this approach is adhered to too religiously, it can lead to some inflexibility in the R&D program. The R&D budget must take corporate needs into account. For example, a decision at the highest levels of management to increase the introduction of new products into the marketplace would probably require restructuring the R&D budget, and last year's budget would provide little guidance for this year's R&D needs.

Base the R&D Budget on Profit Levels

The rationale behind this approach is to let profits pay for R&D by plowing back a certain share of profits into R&D. This approach appears to treat R&D as a luxury which can be indulged in only when times are good. There are many obvious problems with this. For example, those occasions when profits are low (or negative) are the very times when the firm should consider increasing its R&D commitment so that the profit picture can improve. Yet, by pegging R&D expenditures to profits, the firm cuts down on R&D in bad times.

Another problem is that profit levels tend to fluctuate strongly from year to year. Consequently, an R&D budget pegged to profits will also fluctuate dramatically, resulting in a great deal of instability in the R&D program.

Base the R&D Budget on Sales Levels

This is a common way that firms set R&D spending levels. Unlike profits, sales tend to be rather uniform from year to year, so that basing the R&D

Table 4-1
R&D Expenditures for Selected U.S. Industries, 1981

	R&D as Percent of Sales	R&D as Percent of Profits	R&D Dollars per Employee
Aerospace	4.8	141.6	3,717
Appliances	2.0	82.9	1,136
Chemicals	2.5	47.1	2,879
Computers, hardware	5.0	94.6	3,324
Computers, peripherals/services	5.9	94.2	3,284
Drugs	5.3	57.1	4,044
Electronics	3.1	74.3	1,829
Food & beverages	0.7	16.3	761
Fuel	0.5	8.9	2,287
Instruments	4.6	87.5	2,571
Metals & mining	1.1	21.8	1,239
Paper	0.9	16.4	930
Semiconductors	7.1	174.0	3,109
Steel	0.6	10.7	577
Telecommunications	1.2	10.9	770
Textiles, apparel	0.4	19.0	237

Source: "R&D Scoreboard 1981," *Business Week,* 5 July 1982, pp. 55-74.

budget on sales adds stability to the R&D program. Furthermore, data on the percentage of sales devoted by firms and industries to R&D are commonly available, so that by using this measure a firm has a good idea of where it stands in relation to its competitors.

Table 4-1 provides some data on R&D expenditures by selected American industries. As the table makes clear, spending levels vary dramatically from industry to industry.

Notes

1. A.F. Ehrbar, "United Technologies' Master Plan," *Fortune*, 22 September 1980, p. 98.

2. *Internal Revenue Code*, Section 174.

3. L. Smith, "The Lures and Limits of Innovation: 3M," *Fortune*, 20 October 1980, pp. 84–94.

4. Ehrbar, "Master Plan," p. 98.

5. J. Watson, *The Double Helix* (New York: Atheneum, 1968).

6. E.B. Roberts, "New Ventures for Corporate Growth," *Harvard Business Review*, 58 (July–August 1980): 134–142.

7. "R&D Scoreboard 1981," *Business Week*, 5 July 1982, p. 73.

5 Technology Transfer: Mechanisms

Introduction

A major element of international business is the transfer of technology from one country to another. Sometimes the explicit objective of an international transaction is technology transfer, as when a German firm purchases a license to use an American firm's metallurgical process in Europe. Other times technology transfer is simply a side-effect of the purchase of a good. For example, when an American consumer purchases a French Cuisinart food processor for use in his kitchen, he is also purchasing the technology embodied in the device. In both of these examples, technology is being transferred across national boundaries.

In this chapter, we will discuss some of the basic concepts of technology transfer. The discussion will be relatively brief, since an enormous amount has been written about technology transfer already. Attention in this chapter focuses on mechanisms of technology transfer. The next chapter will be concerned with the role played by multinational corporations as agents of technology transfer.

Transfer of technology is an imprecise term that takes on many meanings according to the context in which it is used. The multiple character of technology transfer can be seen by examining some of its varied dimensions. Four dimensions will be treated here.

The international versus domestic dimension. While the term technology transfer is most commonly used to describe international transactions, it applies to domestic situations as well. Domestic firms are constantly engaging in technology transfer when they purchase products from each other; license each other's technology; trade patents; examine competitors' products; and so on. Very often, governments work hard to encourage the acceleration of technology transfer domestically. They do this most effectively by increasing industry's awareness of government-funded or developed technologies that might have commercial applications. For example, in the United States the National Aeronautics and Space Administration (NASA) engages in a major effort to promote the commercialization of space technology through its Technology Transfer Division.[1] On a broader scale, research reports produced as a consequence of non-classified government-supported research efforts are made available to the public at cost through the National

Technical Information Service (NTIS), while, prior to the budget cutbacks of the early 1980s, descriptions of federally funded on-going research projects could be obtained from the Smithsonian Science Information Exchange (SSIE).

The commercial versus non-commercial dimension. Technology transfer can occur at both the commercial and non-commercial levels. Commercial transactions entail the purchase of tangible goods and intangible know-how. Non-commercial technology-transfer generally occurs in the public-goods area, and is promoted by both governmental and non-governmental organizations. It is commonplace in public-health activities. A dramatic example of this was the successful effort by World Health Organization (WHO) personnel to eradicate an age-old scourge of mankind—smallpox. This was done by training public-health personnel throughout the world to cope effectively with local outbreaks of the disease. Non-commercial technology-transfer is also commonplace in agriculture. Domestically, farmers are made aware of the latest advances of agricultural technology through agricultural extension services. Internationally, agricultural and nutritional know-how is transferred to developing countries under the auspices of national foreign-aid programs or programs sponsored by international organizations, such as the United Nations Food and Agriculture Organization (FAO).

The tangible goods and processes versus intangible know-how dimension. Technology transfer can involve the transfer of tangible goods and processes. Examples of this are the construction of turn-key operations in foreign countries, the acquisition of naked technology, and the purchase of off-the-shelf items. It can also involve the transfer of intangible know-how. This is exemplified by on-the-job training programs, educational exchange activities, and information gleaned from scientific and engineering journal articles.

The free versus proprietary knowledge dimension. Finally, technology transfer can entail the acquisition of free knowledge or proprietary knowledge. Free knowledge is often accessible in publications appearing in the public domain, such as scientific journal articles and government reports. In general, scientific knowledge can be obtained quite freely, since one of the fundamental norms of the scientific community is the free and open promulgation of research findings.[2] In contrast, technological knowledge is often proprietary and may be protected by patents or trade secrets. If the proprietor of a given piece of technical knowledge is willing to release this knowledge, he will generally do so for a price.

Since technology covers so much ground, any definition of it must necessarily be broad. The definition we offer is this. Technology transfer is the conveyance of either a man-made tangible good/process or intangible know-how from those who possess it to those who do not.

Basic Elements of the Technology Transfer Process

Although technology transfer encompasses a great many different types of activities, the transfer process itself can be described in rather simple terms. Any technology transfer process can be examined according to six elements:

Element 1: The transfer item. What is being transferred? A tangible product (for example, a heat sensor)? A process (for example, a semiconductor doping process)? Know-how (for example, information on proper fertilizer use)? Some combination of these things?

Element 2: The technology donor. Who possesses ownership rights to the transfer item? Is he willing to make the technology accessible to others?

Element 3: The technology recipient. Who will receive the transfer item from the technology donor?

Element 4: The transfer mechanism. Precisely how will the transfer item be delivered from the technology donor to the recipient? Through licensing? Direct purchase? Some other mechanism?

Element 5: The rate of diffusion of technology. We can examine the rate of diffusion of technology from two perspectives.[3] First, after an innovating country begins producing a new good, how long does it take for other countries to obtain the production capabilities to produce the good? This is called the imitation lag. Second, after the good first hits the market in the innovating country, how long does it take for it to gain acceptance in foreign markets? This is called the demand lag.

Element 6: The absorptive capacity of the recipient. How capable is the recipient in adopting the technology effectively? Developing countries, with their weak scientific and technological infrastructures, have particularly poor capacities to absorb new technology. Consequently, their choice in selecting technologies suitable for their development needs is severely limited.

Transfer Mechanisms

There are innumerable ways in which technology can be transferred from donor to recipient. Licensing alone can be accomplished in a myriad of different ways. We will here look at thirteen of the most frequently used mechanisms for technology transfer.

Turnkey Operations

Typically, in this kind of operation the technology donor constructs a fully functioning production facility for the technology recipient. Once it is completed, the recipient need merely turn a key to get the facility functioning.

Varying degrees of know-how are transferred via the turnkey operation. On the one hand, the recipient can contract to have its technicians and managers thoroughly trained by the donor in the operation of the facility. By doing this the recipient can maintain a modicum of self-sufficiency in running the facility. On the other hand, because of a low capacity to handle the technology emobodied in the facility, the recipient may decide from the outset to depend heavily upon the donor to run and maintain the facility.

One major disadvantage to acquiring a turnkey operation is cost. The purchase of a turnkey operation from a foreign source may mean the loss of critical amounts of hard currency, a particularly high price to pay for a country with scarce hard currency reserves. Another disadvantage is the minimal acquisition of know-how. Since the recipient is buying a finished product (that is, a functioning production facility) it may not acquire many of the engineering and management skills needed to construct the facility.

There are a number of advantages associated with the purchase of a turn-key operation. First, if the recipient is careful in selecting who it will choose to produce the turnkey facility, it will be able to acquire tested, competitive, high quality production facilities that may incorporate the most advanced state-of-the-art technology. Second, the lack of skilled labor in the recipient country need not prevent it from obtaining modern facilities, since much of the construction of the facilities requiring skilled labor will be managed and performed by the technology donor's personnel. Finally, once the facility is functioning properly, it will generate income to the recipient country and contribute to the growth of the economy. However, it is quite obvious that if the recipient elects to purchase a facility whose product does not sell adequately in either domestic or international markets, then the turnkey venture could easily turn into a financial calamity.

The Technological Enclave

Fortunately, the technological enclave in its most egregious form has gone the way of the carrier pigeon. At its worst, the enclave was a fully functioning production facility that was divorced from its surroundings. Typically, it would be a modern outpost of a multinational corporation operating in a less-developed country. Aside from employing some of the local population as laborers, the enclave would do little to ameliorate the economic lot of the recipient country. Since the management of the facility would be composed of foreigners on the donor's staff, very little know-how would be transmitted to the indigenous population. Quite often, the products of the facility would not find their way into the local economy, but would be exported to foreign markets.

Technology enclaves flourished in the post-World War II era through the 1960s. Developed country companies would set up enclaves in developing countries for a number of reasons: to be closer to sources of raw material, to take advantage of cheap labor, to have production facilities nearer to foreign markets, or to take advantage of economic incentives offered by the host country (for example, tax holidays). These enclaves were viewed by Third World intellectuals as physical manifestations of neo-colonial exploitation. They became highly visible targets of LDC wrath toward the developed countries.

By the 1970s, most of the corporations that owned the enclaves made conscious efforts to better integrate their overseas facilities into local economies. For example, skilled labor and management positions, formerly reserved for developed country employees, were now opened up to the indigenous population. Interestingly, in the 1970s a new type of technology enclave developed with the active encouragement of certain host LDCs. Countries such as Malaysia, Singapore, and Taiwan promoted the establishment of so-called off-shore production facilities, frequently in the electronics area. These facilities would be owned by multinationals who wished to take advantage of cheap, reliable labor in the host countries. Unlike earlier technology enclaves, however, the host governments would establish ground rules that assured substantial technology-transfer to indigenous workers.

Licensing

One of the chief mechanisms unaffiliated firms employ for transferring technology among themselves is licensing. A license is an agreement that allows a technological recipient to employ a donor's technology, subject to certain conditions that are spelled out by the parties to the agreement. Usually, it is agreed that the recipient will pay a fee or royalty payment to the donor for use of the technology. Often the royalty payment is a small percentage (for example, five percent) of the recipient's sales that were made possible through use of the donor's technology.

This technology transfer mechanism is most useful in situations where the technology recipient has a high capacity to absorb the technology being transferred.

One benefit of licensing someone else's technology is that it reduces duplication of effort and can save considerably on the costs of in-house research and development. The skillful employment of licensing as a strategy for national development enabled Japan to build its economy to a great extent on technology that originated outside its borders. It has been noted that for $3 billion post-World War II Japan acquired the best technology the West had to offer. In view of the enormous earnings gained by the Japanese from this technology, its purchase has to rank high on the list of the great bargains of history.[4]

The terms of specific licensing agreements can vary greatly. The exact terms are arrived at through bargaining between the donor and recipient, and reflect the strengths and needs of both parties. A license may simply involve the transfer to the recipient of the donor's patent rights and engineering data, and may include arrangements to train recipient technicians by the donor's personnel. Or the license may be more far-reaching and include, above and beyond technical assistance, trademark rights and provisions requiring the donor to buy back a certain fraction of the good produced by the licensee.

Licensing will be discussed in greater detail in chapter 7, along with the two topics that are treated next—joint ventures and patents.

Joint Ventures

A joint venture occurs when two or more business entities set up a third entity that will enable them to produce a good or service jointly. One common motivation for undertaking a joint venture is to share costs. This motivation is particularly evident in projects that involve enormous expenses, such as major aerospace and computer development initiatives. The first case is illustrated by the Anglo-French venture to build a commercial supersonic transport aircraft (the Concorde), the second by the ill-fated French-German-Dutch computer consortium—called Unidata—that was designed to mount a coordinated attack on IBM. Another common motivation for pursuing a joint venture is the desire to share technical, marketing, production, and managerial skills. For example, an American firm may have technical and marketing expertise, but may be weak in production, while a Korean firm may be strong in production (owing to its comparative advantage in labor costs), but weak in other areas. By working together, the American and Korean firms can build upon each other's strengths. Generally, joint ventures involve both cost-sharing and skill-sharing arrangements.

The actual process of sharing technology in a joint venture can occur in a number of ways. For example, one partner may issue a license to another to use its technology freely in the production of a good. If the technology is of high proprietary value, however, its owner may choose to maintain complete control over its manaufacture and simply supply the crucial technology to the production partner in a packaged form, allowing the production partner to incorporate it into the final product. If both parties have strong technological capabilities, technology sharing might be undertaken by means of a cross-licensing arrangement, where each partner can have access to the other's technology.

Patents

A patent gives an inventor the legal right to possess monopolistic control over his invention for a stipulated term. In the United States, this term is seventeen years. A patent protects an invention only in the country in which it is issued. If a firm wishes to obtain patent protection in more than one country, it must apply for the additional protection in each country within a year of the first patent application.

Like other types of property, patents can be bought and sold. That is, patent rights are transferable. There are a number of reasons why one firm might want to purchase another firm's patents. For example, it may wish to buy a patent for a technology that will help it to round out an existing product line. Or else, it may buy a patent in order to get into a new line of business. Or else, it may purchase a patent in order to avoid a patent infringement suit.

The transfer of patent rights from one party to another may involve payment other than cash. Patents can be swapped between companies, transferred in exchange for an equity holding in a company, and so forth.

Clearly, the purchaser of a patent needs to have a good technological absorptive capacity if the acquired patent rights are to be used effectively.

In-house Transfers to Foreign Subsidiaries

Multinational corporations (MNCs) are the principal agents of international technology transfer in the world today. Because detailed records of their business transactions are withheld from public view, it is hard to measure just how much technology has been transferred between countries via the MNCs. One crude measure of MNC technology transfer is published data on royalties and fees paid and received by corporations in their dealings with foreign subsidiaries. These royalties and fees are payments for licenses, copyrights, and trademarks. The figures are substantial. For example, it is estimated that in 1981 U.S.-based companies received $5.9 billion in royalties and fees for direct investment abroad, and paid out $516 million.[5]

Regrettably, these data have many problems associated with them. For example, some of the data do not reflect the true value of transferred technology, since the internal pricing of a technology may simply be a means used by corporations to transfer profits out of the host country. In addition, most payments are spread out over time, so that the figures include multiyear as well as annual payments. Nonetheless, despite the uncertainties associated with these data, they suggest in a rough way that in-house technology transfers by corporations to their foreign subsidiaries are considerable.

It is not surprising that multinational corporations are major technology-transfer agents when one considers their operating procedures. First, MNCs function largely without regard to national borders. Resources are allocated to locales where they can be employed most efficiently. If the economics of a product's development so dictate, the product may be designed in country A, manufactured in country B, and packaged in country C.

Second, MNCs view subsidiaries as partners rather than competitors. The proprietary concerns a company may have in dealing with an unaffiliated firm are largely absent when dealing with a subsidiary. The combination of these two factors makes MNCs efficient and large-scale technology-transfer agents.

Some concern has been expressed by the MNCs' home countries regarding the MNCs' role in exporting technology abroad. The concern focuses on the fact that other countries may obtain at little cost the fruits of expensive R&D efforts undertaken in the home country. Ultimately, such technology policies can strengthen the competitive position of the recipient countries in world markets and result in a weakening of the donor country's trade position.

Simple Emulation of a Product or Process

One obvious way to obtain technology developed elsewhere is simply to copy it. If the technology is reasonably complex, this will require strong scientific and technological capabilities on the part of the technology recipient. A common way to copy a technology is through reverse engineering, where the finished product is carefully examined and taken apart to see what makes it tick. A more pernicious way to copy a technology is by means of industrial espionage, where the design of the technology is obtained illegally through subterfuge.

The emulation of other firms' technology need not be an entirely negative development. It is sometimes argued that in a fast developing technological area, reverse engineering can stimulate innovative activity. For one thing, if a technology can be copied rapidly, then the best way for an innovative company to stay ahead of emulators is to innovate continuously. For another, copying a firm's R&D efforts reduces duplication of effort and enables the copying firm to focus its research efforts on innovations in other areas.

One industry where the copying of technology is widely practiced is the semiconductor industry. Recent attempts to extend U.S. copyright protection to semiconductor circuits failed, in part, because many members of the semiconductor industry were reluctant to give up their reverse engineering practices.[6] While copying the works of competitors is viewed as stimulating innovation in the semiconductor industry, it must be acknowledged that the practice penalizes the most innovative firms, whose R&D exenditures support not only their own sales but the sales of competitors as well.

Direct Purchase of Naked Technology

If a firm wishes to manufacture a product that is based on a complex technological device, it may elect to purchase it outright from another firm rather than manufacture it itself. Such a purchase of naked technology can save the technology recipient considerable R&D and production costs.

It has been argued that American firms have been too ready to sell naked technology to foreign firms.[7] As a consequence, the foreign firms are able to benefit rather cheaply from costly R&D efforts undertaken in the United States. Furthermore, easy access to state-of-the-art technology may enable them to manufacture products that are more competitive in international markets than American products, hurting the U.S. trade position as a result.

Purchase of Embodied Technology

From the point of view of the donor country's balance of trade, the sale of a key technology embodied in a product is generally preferred over the sale of the same technology in naked form. For example, a special semiconductor device may be the key element of an instrument that measures high temperatures. Should the donor sell only the semiconductor device to a foreign recipient, it will earn less per unit sold than if it sold the whole instrument. Furthermore, by selling the device alone, in naked form, donors risk having foreign manufacturers incorporate it into an instrument of their own making that could erode the donor's competitive position in world markets.

Purchase of Technological Services

When a country lacks certain scientific and technological skills, it can remedy this deficiency by hiring technological services from abroad. There are many ways in which this can be done. For example, a large American company working with enamel products hires enamelists from Europe, where a sufficient supply of these skilled individuals exist. European oil exploration companies sinking wells into the North Sea hire experienced American oil technicians from the Southwest United States to help them in setting up oil rigs. Smaller countries that are rapidly industrializing hire foreign engineering firms to design and build state-of-the-art industrial plants.

As with a number of the other technology-transfer mechanisms, the extent to which technology is transferred from donor to recipient is highly variable here. When a country with low science and technology capabilities hires foreign teachers at the primary, secondary, and university level, it is making an important investment that will have high pay-offs in the long run. Enormous amounts of valuable information can be transferred from donor to recipient through a decent educational system. If it is sufficiently extensive, the educational process will radically transform the society and outlook of the recipient country, making it more amenable to the indigenous rooting of science and technology.

On the other hand, a country that hires foreign technical personnel to perform duties that its own population cannot handle may find that little or no technology is transferred to its citizens, unless explicit provisions are made to train them. The simple purchase of other people's skills may postpone the day when the country can exercize a modicum of self-sufficiency in a highly technological world.

Education Abroad

Historically, education abroad has been a very important technology-transfer mechanism. It was certainly important in the development of scientific and technological capabilities in the United States during the eighteenth and nineteenth centuries. At that time, Americans lacked first-rate domestic institutions of higher education and received their technical training abroad—primarily in England, France, and Germany. Education abroad is no less important to developing countries today.

The major training ground for students studying abroad is the United States. Some 300,000 foreign students are enrolled in American educational programs in any given year.[8] More than half of them are receiving training in science and engineering.

The principal drawback with education abroad as a technology-transfer mechanism is its cost and long gestation period. The costs of education take many forms. There are the direct costs of supporting a student's living expenses, tuition, travel expenses, and so on. There are also substantial opportunity costs. For example, while the student is studying abroad, he is not generating any income or making an immediate contribution to society. Additionally, money spent on his education is not being put to other uses that may have a more direct bearing on a country's problems.

Despite these limitations, a country wishing to develop adequate indigenous scientific and technological capabilities has no choice but to send students abroad for education.

Site Visits and On-the-Job Training Abroad

Sending people abroad simply to observe how others operate a piece of machinery, or manage a complex process, or utilize resources can be a very cost-effective way of transferring technology. Of course, it is important that the individual on a site visit, or participating in on-the-job training, have some prior idea of what it is that he is supposed to learn. The greater his competence in his specialty area, the greater the amount of technology that will be transferred to him.

International Cooperative Research Efforts

In the technologically advanced countries, programs have been set up that encourage international research collaboration. These programs may take several different forms, ranging from making funds available for individuals to attend international conferences to supporting an international nuclear research center (the European CERN). These endeavors generally focus more on transferring scientific information than on transferring technology.

Published Literature

One of the principal norms of science is to disseminate research findings. This is done mainly through scientific journals. It can reasonably be said that the journal literature is the repository of virtually all scientific knowledge. Consequently, the published literature is an excellent—and relatively cheap—source of information for scientists. Any scientist who does not regularly scan the journals in his discipline will soon be out of touch with recent developments.

Probably the principal deficiency of the published literature as a mechanism for transferring scientific information is that by the time the information is finally published it may be out of date. First, time lags are introduced when the researcher writes up his findings. The time lag lengthens after the article is sent to a journal and is circulated among reviewers. Frequently, the comments of the reviewers require a rewrite of the findings. Then, of course, more time passes as the findings are type set and the journal is sent out to libraries and other subscribers. It is not at all uncommon for findings to appear in a journal several months to two years after the initial discovery. In a fast-paced area, this may mean that even as an article is rolling off the presses, the findings it reports are obsolete. The matter of delays is even more significant in the case of LDCs who receive journals from the developed countries by means of surface mail. The mail transit alone may consume several months.

Meetings, Seminars, and Colloquia

The delays inherent in publishing scientific findings are partly compensated for when scientists gather together. At these gatherings they are able to report on research in progress as well as completed research. A great deal of the information transfer does not occur duing the delivery of scientific papers, but rather occurs informally in the corridors of the conference facilities.

This is an excellent way of disseminating scientific information in a timely fashion, provided, of course, that funds are available to send scientists to international gatherings.

Conclusions

It is clear that technology transfer can occur in a variety of different ways. It also should be clear that there are no technology transfer mechanisms that are universally best. Purchasing turnkey operations may be appropriate for a rich LDC such as Saudi Arabia, but singularly inappropriate for impoverished Chad. Licensing may be a good technology-transfer mechanism for South Korea, with its solid technology base, but not viable for little Surinam. The appropriateness of a particular transfer mechanism is defined by the specific circumstances in which a country finds itself.

Notes

1. See, for example, National Aeronautics and Space Administration, *Spinoff 1981* (Washington, D.C., 1982).

2. This is reflected, for example, in Merton's assertion that scientists have developed "a passion for eponymity rather than anonymity." Robert K. Merton, *The Sociology of Science* (Chicago: University of Chicago Press, 1973), p. 33.

3. J.E. Tilton, *International Diffusion of Technology: The Case of Semiconductors* (Washington, D.C.: Brookings Institution, 1971), pp. 19-47.

4. R. Goldscheider, *1979 Technology Management Handbook* (New York: Clark Boardman Co., 1979), p. 4.

5. Bureau of Economic Analysis, *Survey of Current Business* 62 (March 1982):58-63.

6. "How 'Silicon spies' Get Away with Copying," *Business Week*, 21 April 1980, p. 180ff.

7. See, for example, M. Boretsky, "Trends in U.S. Technology: A Political Economist's View," *American Scientist* 63 (1975):70-82; and S. Gee, "Foreign Technology and the United States Economy," *Science* 187 (1975):622-626.

8. National Science Board, *Science Indicators 1980* (Washington, D.C., 1982), p. 240.

6

Technology Transfer: The Multinational Corporations

MNCs and Technology Transfer

Multinational corporations have been defined in numerous ways. Definitions range from the very broad to the narrow. At one end of the spectrum, they can be defined as firms that function in two or more countries. This is a rather loose approach and includes small mom-and-pop operations that happen to do business in two countries as well as giant corporations whose plants dot a map of the world. At the other end of the spectrum is a commonly used and rather restrictive definition of an MNC as a firm that has annual sales of over $100 million and operates in six or more countries. The common view of the MNC, that it is a mammoth organization with tentacles reaching around the globe, is closer to the second definition than the first. The second definition includes only a few hundred companies—about one-half of which are American based—while the first includes thousands.

MNCs are often engaged in manufacturing, although many well-known MNCs are also involved in mining, banking, and the recovery, refining, and distribution of petroleum. The manufacturing fields of the firms are frequently fast moving, technologically-based areas, such as computers, electronic equipment, and pharmaceuticals. Consequently, R&D plays an important role in the operation of many MNCs. The R&D budgets of a number of them are enormous. For example, in 1981 the R&D budget was $2.250 billion for General Motors, $1.718 billion for Ford, $1.612 billion for IBM, $1.686 billion for AT&T, $844 million for Boeing, $814 million for General Electric, $736 million for United Technologies, $615 million for Eastman Kodak, $631 million for DuPont, and $630 million for Exxon.[1]

MNCs can function very efficiently on a global scale. This efficiency is largely based on the fact that corporate activities are controlled in a central office that is aware of economic conditions and corporate undertakings throughout the world and can take advantage of opportunities where they occur, rapidly shifting resources and responsibilities from locations where they are not employed effectively to places where they can be. For example, an MNC may decide to set up some of its operations in a country that extends good investment terms to it (for example, a tax holiday for the first five years of operation). If this country has relatively cheap labor, the local facility may focus on labor intensive assembly operations for a product. Meanwhile, automated manufacturing operations may be carried out in another country, which has a comparative advantage in automated production. Finally, sales of the completed product may be directed at third countries.

Because of the global nature of their activities, MNCs can be very efficient technology-transfer agents, particularly in their dealings with wholly owned subsidiaries. Normally, when technology is transferred between two unaffiliated parties, tedious negotiations must be undertaken that govern the conditions for the transfer. The technology donor may want to make sure that the recipient protects it from third parties, in addition to arranging terms of payment for the technology. The technology recipient may want to make provisions guaranteeing that the technology is updated continuously, assuring that the engineering specifications are complete, arranging for training at the donor's facilities, et cetera. The negotiations can be protracted and costly, and ultimately one or the other party may be dissatisfied with how things turn out. When dealing with a wholly owned subsidiary, these problems are reduced. Headquarters may determine that a new production process invented in its New York laboratories should be adopted immediately in its plants in Illinois, Mexico, and Great Britain and the transfer of technology can be effected in the minimum amount of time.

Technology transfer can occur at different levels. On the one hand, the technology transfer can be so thorough that the recipient masters the technology to the point that he can replicate it in its entirety and can even make improvements on it. On the other hand, it may be that virtually no knowledge or skills are transferred to the recipient, as when he purchases a product off-the-shelf that embodies a technology that is incomprehensible to him. The degree to which the transfer is thorough depends upon factors such as:

1. *The absorptive capacity of the recipient.* If he has the technical capabilities to understand the imported technology, he is in a good position to insist that the donor show him how to employ and nurture it. In fact, if his technical skills are very good, he may be able to understand the technology through reverse engineering and avoid dealing with the donor. The reverse engineering approach is a common technology acquisition technique employed by the Soviet Union.

2. *The recipient's bargaining skills.* If the recipient is thoroughly conversant with the target and related technologies, the donor's position regarding sales of its technology, et cetera, his bargaining position can be considerably improved. He will have a far better idea of how much he can reasonably demand from the technology donor than the recipient who has not done his homework, and as a result he is more likely to strike a satisfactory deal.

3. *The recipient's purchasing clout.* The recipient who has ample hard currency reserves available to him for the purchase of technology is in a far

better position to obtain concessions on substantial transfers of know-how than the recipient with a shaky credit standing who requires long-term financing in order to purchase the technology.

4. *The recipient's market.* If the technology recipient operates in a market where there is a strong demand for his products, assuring him of strong sales, this fact can be used to extract concessions from the donor. This is especially true in the case where the technology is being transferred through a licensing agreement, since royalties to the donor are generally based on sales volume.

5. *The position of the donor's competitiors.* The donor who has competitors who are able to supply technology that is similar to its own is more likely to make substantive technology-transfer concessions than the one who has complete monopolistic control over its technology.

As it turns out, firms in developed countries tend to score highly on the above factors, while LDC enterprises do not. Operations in developed countries have high absorptive capacities, good bargaining skills, access to good financing, and good markets. Consequently, technology-transfers from MNC donors to enterprises in developed countries tend to entail more thorough transfers, while those to LDCs tend to be rather superficial. Thus, in looking at MNCs as technology-transfer agents, it is useful to distinguish between their dealings with developed countries and LDCs.

MNCs and Technology Transfer to Developed Countries

In the 1950s and 1960s, when MNCs (which were primarily American based) dramatically expanded their international activities, they generally operated through foreign subsidiaries over which they had substantial control. In these years, the great bulk of international business was undertaken among the developed countries of the West. As long as business was conducted through wholly owned subsidiaries, the MNCs were able to maintain control over the technology they had developed. Technology transfer was thus kept at a fairly superficial level. Transfers occurred principally through the training of host country personnel in American management and production methods and through the sale of products that embodied proprietary technology. Key technologies, however, were maintained as proprietary knowledge and for the most part were not transferred to the host countries.

In the 1970s, several events came together which caused the MNCs to loosen their grip on their proprietary technology. First, economic conditions forced many of them to pull back from their overseas operations. These conditions included the weakening of the dollar, which increased the cost to U.S. firms of doing business abroad; unsatisfactory earnings from overseas operations; a change in the income tax law which reduced the deductions allowed Americans working overseas; recession and inflation at home and abroad; and the feeling that protectionism was on the upswing globally. Second, increased international competition enabled potential technology-recipients to shop around. What was clearly a seller's market in the 1950s was turning into a buyer's market. With increased competition, buyers were able to demand increased access to proprietary technology. Third, governments were injecting themselves increasingly into foreign-investment decisions; either overtly, as in the case of the Eastern European countries not allowing equity ownership of firms on their territory; or more indirectly, as in the case of the French government financially subsidizing domestic computer firms, or the Japanese government procuring telecommunications equipment only from domestic firms.

As the 1970s progressed, protectionist sentiments grew in the developed countries and arms'-length licensing agreements and joint ventures increased. Both these mechanisms can lead to significant transfers of technology. For example, the technical assistance portion of a license agreement might call for complete access by the technology recipient to the donor's patents; engineering data; information on advanced production techniques; and continuous updates on improvements in the technology. In addition, the licensing agreement may contain provisions for training personnel at the donor's and recipient's facilities. Joint ventures can also entail substantial transfers of technology, particularly if they involve the coproduction of goods. With coproduction, the donor shares its technology for production purposes with its foreign partner.

Clearly, licensing agreements and joint ventures can require the donor MNC to release substantial quantitites of technology to unaffiliated firms. This fact has led to criticism, particularly in the United States, that MNCs are giving away valuable national assets and contributing to the country's competitive decline in world markets. Alarm regarding such give aways is particularly great in the case of U.S. dealings with Japan. While erecting various barriers protecting indigenous industry from foreign business incursions, the Japanese have built a powerful domestic industry largely through the licensing of foreign technology. Consider that between 1966 and 1981 unaffiliated Japanese firms paid American firms some $3.8 billion in royalties and fees, while unaffiliated U.S. firms paid the Japanese only $165 million, a ratio of 23 to 1![2] To be sure, the donor MNC can attempt to protect its technology. One approach calls for the measured release of core technology, where the donor supplies the recipient with a key technology ready-made and withholds information on how it is produced. Another approach is to include provisions in the joint venture or license agreement that prohibit the recipient from sharing its technological know-how with third parties. Even when attempts are made to control others' access to proprietary technology, substantial transfers occur, if not of the key technology, then of ancillary technologies.

MNCs and Technology Transfer to LDCs

It is not as easy to generalize about LDCs today as it once was. At one time, virtually all LDCs had the following attributes in common: they had low per-capita incomes, low educational levels, poor nutrition and health, and very little industrial capacity. Perhaps one of the few features that distinguished some of them from others was their *potential* for growth. That is, the socioeconomic systems of a number of the LDCs were more conducive to fostering rapid development than the socioeconomic systems of others. In the decade of the 1970s, it became apparent that the high growth potential of some countries was creating a number of development success stories. In Asia, what are called the four little dragons—Korea, Taiwan, Singapore, and Hong Kong—consistently realized high growth rates (over nine percent each year in real terms, compared with six percent for Japan and three percent for the United States).[3] In Latin America, Brazil and Mexico showed great prom-ise, while in Black Africa the shining star was Nigeria. A number of countries realized enormous growth in their economies not because they pursued par-ticularly brilliant growth strategies, but simply because they happened to have the good fortune of being located above massive petroleum reserves.

In order to appreciate the great social and economic changes that have occurred in the world since World War II, it should be noted that by a num-ber of economic measures (for example, per capita income), Japan could have been regarded as an LDC in 1950. Of course, Japan differed from to-day's LDCs in that she had substantial industrial capacity, as was demon-strated by her ability to sustain a major war effort between 1941 and 1945.

In view of the substantial differences among LDCs today, it should be recognized in the ensuing discussion about the MNCs' dealings with LDCs that our generalizations are not as appropriate today as they were ten or twenty years ago.

Unlike the advanced countries, most LDCs, particularly the poorest among them, are not in a position to extract many significant concessions from MNCs on the issue of access to proprietary technology. As was men-tioned earlier, they have a low science and technology absorptive capacity; they are at a disadvantage when bargaining with MNCs, owing to their lack of scientific and technological skills and knowledge; unless they are blessed with resource riches (for example, petroleum), they lack purchasing clout; and their domestic markets are generally weak. Indeed, many LDCs have a hard enough time attracting MNCs to their countries under any conditions, and bargaining over access to technology, while not the least of their prob-lems, is not their greatest worry.

LDC attitudes toward MNCs are highly ambiguous. On the one hand, MNC investments are welcome because they bring with them much needed capital, create jobs and income, introduce technology into the economic system, train people in useful skills, and contribute to industrialization. On the other hand, MNCs are viewed with great suspicion and distrust. There are a number of reasons for this. First, an LDC may fear the great economic power of an MNC, whose global assets may be substantially greater than the country's GNP. Second, LDCs are reluctant to surrender some of their national autonomy to an outsider. To many inhabitants of LDCs, the differences between the old colonial relationships and the new ties with MNCs are very small. In a related vein, a significant number of people in LDCs see the MNCs as agents of foreign governments who act to further the foreign-policy objectives of these governments. They point to the case of United Fruit's operations in Central America as evidence. Finally, LDCs recognize that the MNCs' interests differ from their own national interests. While the LDC may want an MNC to play an important role in its development plan, the MNC itself is operating according to corporate objectives that may or may not correspond to the LDC's development aims.

Because so many LDCs occupy a weak position in their demands for technology, transfers to them are apt to be more superficial than to already developed countries or to LDCs who have graduated into the category of Newly Industrialized Countries (NICs). This is possibly best illustrated by reference to off-shore sourcing. With off-shore sourcing, firms in advanced countries set up production facilities in LDCs, where labor is cheap for production and assembly tasks. The manufactured and assembled products are then sent back to the home country, where they are marketed for domestic consumption, or else they may be sent directly to foreign markets. The best known example of off-shore sourcing is the case of U.S. electronics firms having their products manufactured in several Asian countries. These firms obtain a number of benefits in off-shore sourcing beyond the obvious benefit of cheap labor. Additional benefits, particularly in the Asian cases, include little labor-unrest and high worker-discipline. Furthermore, the advanced country firms can use their overseas operations in ways that would be untenable at home. For example, the overseas facilities may be employed to meet orders in boom times. Once the boom is over, plants may be closed and indigenous workers laid off until the arrival of the next boom. (Clearly, in such a situation, labor stability is rather poor for the workers at the off-shore plants.)

Off-shore sourcing generally involves low levels of technology transfer. Most tasks performed in the off-shore manufacture of a product are highly automated. The tasks that require human intervention are generally trivial assembly operations that employ unskilled labor, so that the typical worker picks up few useful skills in the plant. To the extent that local management is used to run the plant, they will gain useful exposure to modern production techniques. These production techniques and the advanced technology that goes into making the product are developed in the home country.

Despite the obvious shortcomings of off-shore sourcing agreements to the host LDC, they are fairly attractive to countries that are not preoccupied with the issue of *dependencia*. One interesting appeal of the off-shore sourcing arrangements to the host LDC is the fact that the products being manufactured are designed for export. Consequently, these products will not compete with local manufacturers in domestic markets. Other advantages include employment and income generation, and government revenues from the operation.

Technology Transfer to LDCs via Foreign Direct Investment

Traditionally, foreign direct investments in LDCs yield little high level technology transfer. An important reason for this is that virtually all the R&D for an MNC is done in the home country. There is little incentive to set up R&D facilities in the host LDC. For one thing, the MNC is concerned that ultimately such facilities will work to the firm's disadvantage, since they may improve the host LDC's bargaining position on technological matters, may improve the LDC's ability to imitate the MNC's technology, and may enable the host LDC to develop its own national technology. Beyond this, the MNC is protective of the technology it has developed at great expense, technology which may be vital to the firm's international competitive position. It is naturally reluctant to part with it, or even to place it in the hands of employees who are citizens of the host country. This last point is attested to by the fact that the highest proportion of expatriates in the typical host country operation work in top engineering and technical posts.[4] To the extent that LDCs do establish local R&D operations, these operations focus on quality-control tasks and adaptation of MNC technology to local conditions.

The question of whether or not MNCs establish R&D facilities in LDCs is not a trivial one. The presence of MNC laboratories in LDCs could be a major technology-transfer mechanism, particularly if the laboratories perform work that is relevant to the development of the host LDC. These laboratories would be important for the development of indigenous R&D capabilities in a number of ways. First, they would provide some LDC personnel who have earned advanced degrees in science or engineering with meaningful work. This would help combat the curse of technological underemployment (and unemployment) that is so common in LDCs and that contributes measurably to the brain drain. MNC laboratories set up in LDCs would also give LDC personnel much needed experience in working in an advanced, well-run industrial laboratory. These personnel may receive first-rate technical training in the world's finest universities, but the gap between university training and practical R&D can be substantial, and without hands-on experience in a good industrial laboratory the university training may be wasted. Second, the MNC laboratories can serve as a window to the rest of the world's scientific and technological activities. One of the greatest obstacles to the fostering of science and technology in the LDCs is the isolation of scientists and engineers from the mainstream. Well-supplied and well-staffed MNC laboratories in LDCs can go a long way in overcoming this problem.

In the absence of MNC R&D facilities in the host country, how can technology get transferred from the MNC to the LDC via the MNC's subsidiary? D. Germidis identifies four ways.[5] First, and most obviously, technology can be transferred by training indigenous personnel to perform their required functions in the local MNC facility. In the early years, training was principally directed at low level personnel who would perform tasks that varied in their usefulness to the LDC. Individuals who received training in the industrial arts (for example, welding) would gain skills that would be very valuable to the LDC. Those who performed highly specialized routine tasks (for example, assembling semiconductor devices) would have far less valuable skills. As time passes, the LDCs might increasingly demand that local personnel be given significant jobs at middle and upper management levels, so that they can build a cadre of local managerial talent. Today, a number of LDCs have employment quotas requiring MNCs to have a certain portion of the management staff in their subsidiaries coming from the host country.

A second way technology can be transferred to the LDCs is by having the MNC contract out research to local R&D facilities in both the public and private sector. In practice, this rarely happens. Local R&D facilities typically have little or no contact with the MNCs.

A third possible way for technology to be transferred to the LDCs is through MNC arrangements with local suppliers. Rather than purchase supplies that have been engineered by firms in the advanced countries, MNCs could show local suppliers how to produce what they require in their local manufacturing efforts. Again, this occurs only infrequently.

Finally, technology can be transferred by the sale in the host country of locally manufactured products that embody technology. Technology transfer will be effected if the MNC educates local users in the use of the product. Or else, the product may stimulate local users and entrepreneurs to identify novel uses for it, and this may lead to local innovative R&D efforts.

Technology Transfer to LDCs
via Unaffiliated Enterprises

Germidis' four points focus on MNC foreign direct investments in LDCs. LDCs, of course, can also acquire technology through licensing agreements and joint ventures. In such situations, they are apt to have better access to higher level technology. The richer the LDC, and the better its indigenous scientific and technological capabilities, the greater advantage it can take of licensing and joint venture arrangements. However, for the poorest LDCs, licensing technology and undertaking joint ventures are frequently alternatives that lie outside their reach. The licensing and joint venture alternatives are discussed in some detail in the next chapter.

MNC Technology Transfer Performance in LDCs

The performance of MNCs in transferring technology to LDCs is often strongly criticized, particularly by denizens of the Third World. When we examine these criticisms closely, we find that MNC performance is typically rated on what they could do if they chose to make an all out effort to help LDCs in their development efforts. By these standards, there is a clear gap between the actual and the possible. MNCs as a rule do not make an all out effort to transfer technology to LDCs. In a later chapter dealing with the North-South dialogue we examine LDC attempts to remedy this perceived deficiency by means of codes of conduct for technology transfer by multinationals.

Whether MNCs have done enough for LDCs is a moral issue that is not going to be resolved by reasoned argument. However, we should not under-rate the transfers that have taken place through MNCs, whether consciously or unconsciously. For example, the criticism is frequently offered that MNCs have transferred few useful modern job skills to the populations of LDCs. This means, in part, that MNCs have focussed their attention on exploiting cheap unskilled labor and have not devoted enough effort to developing competent middle and top level managers.

But consider the fact that MNCs, through their subsidiaries, have helped introduce LDCs to modern industrial operations. The mere presence of their operations has alerted LDC inhabitants to the possibilities of industrialization. Individuals who once worked with technologies developed hundreds or thousands of years ago are now suddenly confronted with the marvels of modern production. The presence of MNC operations has contributed to social changes necessary for coping with the modern age. Workers who have traditionally held a relaxed outlook on the passage of time are suddenly subjected to the discipline of the clock. Those whose recent ancestors grazed livestock over unfenced fields find themselves laboring in a central location under a single roof, sharing facilities with many other workers. Many of the tangible skills LDC workers learned at the MNC work bench may have been quite trivial by Western standards, but some of the intangibles—primarily lessons learned about worker discipline, which is so necessary in an industrial setting—are revolutionary when put into the context of the workers' cultural milieu.

The transformation of a traditional society into an industrial society is a painful experience. It has always been so. Karl Polanyi, in his book, *The Great Transformation,* vividly describes England's move from agrarianism to industrialism, and the experience—with its attendant exploitation and social upheaval—appears to be similar to what we have witnessed in the LDCs these past few decades.[6] Even the antiprogress tumult we see in Islamic Iran today has a historical precursor in nineteenth century England, where the Luddites, feeling threatened by the new industrial technology, roamed the countryside smashing machinery.

MNCs and Technology Transfer to Eastern Europe

One of the dramatic new developments of the 1970s was the move by the Soviet bloc countries of Eastern Europe to establish better relations with Western business enterprises. A major objective of this move was to acquire Western technology. With this technology, the Eastern Europeans hoped to improve their own lagging economic performance and, additionally, to begin producing goods that would be competitive in international markets. This interest in Western technology is partly reflected in statistics on Eastern European payments of royalties and fees to the United States for licenses and trademarks. In 1970, payments amounted to $4 million. Eleven years later, they amounted to $89 million, a twenty-two fold increase.[7]

The main obstacle to the development of economic contact between the Eastern Europeans and Western firms was the Soviet Union, which held tight rein on the domestic and foreign policies of its satellites. To be sure, some Eastern Europeans had attempted to loosen the reins—East Germany in 1953, Hungary in 1956, and Czechoslovakia in 1968—but these attempts were brutally crushed by the Red Army. The only European communist country that developed early economic ties with the West was Yugoslavia, which under the strong leadership of Tito conducted its affairs independently of Moscow.

The early 1970s saw a thaw in the Cold War, with moves afoot on both the U.S. and Soviet sides to pursue a policy of detente. Tensions eased considerably, opening the way for increased East-West commercial contact. While Eastern European countries were interested in obtaining Western products and technology, Western firms were titillated by the prospect of penetrating the untapped markets of East Europe.

Commercial arrangements with East Europe are complicated somewhat by the fact that these countries are communist states with planned economies. Being communist, they hold some antipathy towards private ownership of the means of production. Unlike Western firms, which for the most part are owned by individuals, Eastern European firms are owned by the state. Consequently, business negotiations with these firms are for all intents and purposes negotiations with governments. Much of the flexibility that marks negotiations between two private enterprises is thus lost. Furthermore, since private ownership is downplayed in East Europe, these countries do not allow foreign companies to hold equity interest in their indigenous enterprises, and do not permit the establishment of wholly owned subsidiaries. (An exception to this is Romania, which allows up to 49 percent foreign ownership in certain Romanian enterprises.)

The fact that the Eastern Europeans have planned economies also complicates matters. Because resources are allocated to their production uses several years in advance, Western companies may have to wait a number of years before their transactions can be fitted into national plans. Once again, spontaneity and flexibility in dealing with Eastern Europeans is quite limited by their economic systems.

Eric Hayden has noted that there are fundamental differences between the Soviets and Eastern Europeans in their business dealings with Western firms.[8] The Soviets prefer short-term deals through which they acquire Western technology, goods, and production capabilities, and then terminate the relationship with the Western firms once the terms of the deals have been met. The Eastern Europeans, on the other hand, seek long-term relations through which they maintain close contact with Western firms. They are particularly concerned with obtaining continuous updates on the technology they acquire.

Hayden has identified three models that reasonably describe the kinds of nonequity arrangements the East Europeans make with Western firms. Model A simply entails putting a technical assistance agreement into effect. The agreement assures that the East European enterprise has the engineering data and know-how needed to produce a good. The agreement also has a provision requiring the Western firm to continuously provide information on updates on the technology for the life of the agreement.

Model B extends trademark rights to the Eastern European enterprise in addition to providing it with technical assistance. The Eastern Europeans often strongly desire trademark rights, because with these rights the Western firms are obliged to assure that the goods coming off the Eastern European production lines are of equal quality to the goods they produce themselves. If they do not maintain quality, then flawed products bearing their logo may hurt their reputation.

Model C involves an agreement that provides for technical assistance, trademark rights, and the buying back of the resultant product by the Western firm. Clearly, a buy-back provision can help reduce the risk of launching a new venture, since it assures that a certain share of the production output will have a buyer, helping to underwrite the cost of the venture.

In general, East European enterprises prefer to have a model C agreement. In addition to providing them with needed technology, it gives them some assurance that the quality of the goods they produce is up to international standards, and it guarantees them some sales in the early days of the new venture. The actual agreement that is reached, however, depends on the mutual desires and bargaining skills of both sides. If all the Western firm wants out of the agreement is the ability to earn some management fees, royalties, and returns on the sale of supplies, it may be unwilling to agree to the trademark and buy-back provisions. The trademark provision, after all, may entail a great deal of trouble for the Western firm as it monitors product quality, and the buy-back provision may leave it with a large inventory of goods it neither needs or wants. On the other hand, if the Western firm sees the agreement as a way in which it can obtain greater quantities of its products without engaging in capital expansion, and if it sees the agreement as a foot in the door of the East European market, then it may eagerly accept a model C agreement. For in-between situations, the bargaining skills of both sides may play a significant role in determining which model is chosen.

Some of the great hopes Western firms had in the 1970s for exploiting the untapped markets of Eastern Europe were chilled in the early 1980s. The economic and political difficulties of Poland at this time dramatically illustrated the frail state of a large Eastern European economy. As the country teetered on the brink of financial collapse, the government declared martial law and severely curtailed the liberties that had been blooming in the previous decade. Other Eastern European countries displayed signs of economic distress as well and great fears were felt in certain quarters of the West, which had extended enormous credit to East Europe, that the Eastern Europeans would begin defaulting on their debt obligations.

Conclusions

Technology is often the life blood of an MNC. Either through its own R&D or the R&D of others from whom it licenses technology, the MNC comes to possess technology that forms the basis of the products and processes that it sells and uses. The MNC is often reluctant to part with this technology, particularly if it is unique and puts the MNC in a monopolistic position. Yet the MNC may be willing to part with some or all of this technology if the price is right—that is, if the MNC determines that parting with the technology will do it more good than harm.

As it turns out, the price is right more often with developed countries than with LDCs for reasons that were discussed earlier in this chapter. As a consequence, more high-grade technology-transfers seem to be made among MNCs in the advanced countries than between MNCs and Third World enterprises. Leaders of the Third World are understandably concerned that such a situation does nothing to close the gap between developed and less-developed countries; if anything, it increases the gap. Things are further complicated when we learn that inhabitants of the principal donor countries also have misgivings about transfers to certain highly developed recipients (for example, Japan) that are able to use the technology to the donor's disadvantage.

It is unlikely that the role of MNCs as technology transfer agents in LDCs will ever be resolved to the satisfaction of all parties. All that can be predicted with any certainty is that the LDCs will continue to press for increased access to MNC technology, both in their bilateral dealings with MNCs and in international forums. As to the issue of technological give aways, it is likely that having experimented with arm's-length arrangements with foreign enterprises for one or two decades, MNCs are now coming to realize the true value of their technology. In future technology-sharing agreements this realization will be reflected in higher prices for the technology and tougher bargaining by the donor MNCs.

Notes

1. "A Research Spending Surge Defies Recession," *Business Week* 5 July 1982, p. 54.

2. Calculated from National Science Board, *Science Indicators 1978* (Washington, D.C., 1979), table 1-19, p. 159; and Bureau of Economic Analysis, *Survey of Current Business* 62 (March 1982): table 10, 61.

3. L. Kraar, "Make Way for the New Japans," *Fortune* 10 August 1981, p. 176.

4. D. Germidis, *Transfer of Technology by Multinational Corporations,* Vol. 1 (Paris: OECD, 1977), p. 22.

5. *Ibid.,* p. 14-15.

6. K. Polanyi, *The Great Transformation* (Boston: Beacon Press, 1957).

7. Bureau of Economic Analysis, *Survey,* p. 60.

8. E. Hayden, *Technology Transfer to Eastern Europe: U.S. Corporate Experience* (New York: Praeger, 1976).

7 Patents, Licenses, and Joint Ventures

Introduction

Discoveries made by firms can be viewed as assets that will generate income over a period of time. This income can come from the firm's exclusive use of the discovery, or from its renting out the discovery for others to use. Or it can come from a combination of the two. Clearly, the first approach gives the owner of the discovery great control over it, while control is attenuated in the other cases.

In this chapter, we will examine in some detail how firms protect their proprietary technology through patents, as well as the mechanisms by which they can share their technology with others, thereby increasing the income it generates. Two approaches are commonly used in technology-sharing arrangements: one is the license, the other the joint venture. In both these cases, the owner transfers some of the control over his technology to an outside party.

Patents

A patent is a grant by a government to an inventor excluding others from making, using, or selling his invention for a specified period of time. In the United States protection is offered for seventeen years, in Italy and Japan for fifteen years, in England for fourteen years, and in France and the Netherlands for twenty years. Thus, a patent effectively grants an inventor monopoly powers over his invention for a limited duration. This monopolistic aspect of patenting is captured in the title of the world's first piece of patent legislation: the Statute of Monopolies, which was put into effect by the English Parliament in 1624. This act allowed exclusive rights to an invention to be granted to an inventor for a period of not more than fourteen years.

Why do governments grant monopoly powers to inventors? The U.S. Patent and Trademark Office offers three suggestions in answer to this question.[1] First, by offering patent protection, inventors are assured that they can enjoy the fruits of their labors, thus encouraging them to introduce their inventions into the economy.

A second advantage of offering patent protection to inventors is that it enables society as a whole to benefit through increased inventiveness within the society. The assumption is that the more inventions, the better off everyone is.

Finally, since patents are published and made public, knowledge of the invention is available to all. This leads to at least two implications. First, publication of patents can reduce duplication of inventive effort. If, while reviewing issued patents, a prospective inventor determines that an idea he has is already patented, he will know better than to reinvent the wheel. Second, publication of patents may stimulate further inventions, since it is common practice among inventors to review them for ideas.

The encouragement of inventiveness was considered so important in the newly formed American republic that its Constitution even made specific provisions for protecting the intellectual property rights of individuals. Article 1, Section 8 of the U.S. Constitution states: "Congress shall have the power to promote the progress of science and useful arts by securing for limited times to authors and Inventors the exclusive right to their respective writings and discoveries." One of the first acts passed by Congress was the Patent Act of April 10, 1790. It is interesting to note that the first administrator of the U.S. patent system was no less a personage than Thomas Jefferson. It is also interesting to point out that a major invention was granted patent protection not long after the American patent system was established: On March 14 1794 Eli Whitney was granted a patent for his cotton gin.

Inventors patent their inventions for a number of reasons. Most obviously, they obtain patents in order to secure monopoly rights for their inventions. With these rights granted, they can freely invent new products without great fear that the knowledge of how to make them will be pirated by others. This is the intent of patent law. But patent protection can also inhibit the introduction of new inventions. Companies may obtain patents not because they intend to produce a new good, but because they wish to prevent their competition from doing so. A well-known example of this kind of defensive patenting involved the Xerox Corporation. In 1969-1975, the Federal Trade Commission determined that the Xerox Corporation had a stranglehold on the office copier market because the company had obtained patents on all the crucial aspects of dry paper photocopying. Because of this, competitors could not enter the dry paper photocopier market even though they had the technical capability to do so. In a consent decree, Xerox agreed to license out some of its crucial patented technology for a nominal fee, thus introducing more competition into the photocopier market.

Companies may also be encouraged to patent their products because it facilitates licensing or cross-licensing arrangements. Without the protection offered by patents, companies would be reluctant to let others use their proprietary technology, for fear that it might be pirated.

Finally, a company that holds a patent on a desirable invention will have less difficulty in obtaining capital to develop it, because a patent constitutes a tangible and precise claim to the invention.

Trade secrets are an important alternative to patents. There are a number of problems with patents that may lead an inventor to keep secret the process employed to make his product. One disadvantage of patents is that they grant monopoly rights for a limited duration. When the patent expires, the monopoly rights are lost and competitors can copy the process without fear of patent infringement. Another disadvantage is that they are published and may give competitors technical insights that can be useful to them. A third disadvantage is that patents offer protection only in those countries in which they are valid. In countries where an invention is not patented, there are no legal restrictions on copying it. Finally, patent holders must be constantly on guard against infringements and must be willing and able to bring infringers to court. This can be very expensive, costing $50,000 and up in the United States.

These problems associated with patents are absent in the case of trade secrets. However, trade secrets have limitations as well. First, the value of a trade secret is good only insofar as it remains secret. If an inventor independently develops the key technology protected by a trade secret, then the holder of the trade secret has lost exclusive control over the invention. Second, because maintenance of secrecy is vital, great expense may be incurred in establishing proper security to prevent divulgence. Third, while there are specific procedures that can be undertaken to remedy patent infringements, trade secret protection is amorphous. In the United States, for example, protection is based on common law. Protection is basically offered in those situations where individuals who have a confidential or fiduciary relationship to the trade secret holder wrongfully disclose or improperly use the secret.

If trade secrets can be maintained securely, their value may be very great, particularly because they can earn income long after their patentable life would have expired. The formula for Coca Cola is probably the world's most famous trade secret. Other notable trade secrets are the four hundred year-old formula for Chartreuse liquer and the metallurgical procedure discovered by the Zildjiam family for producing the world's finest musical cymbals, a secret that has been guarded closely since 1623![2]

Patenting in the United States

The United States has the single most significant national market in the world. U.S. patents are highly sought after because individuals and firms would like to have monopoly powers over their products in the lucrative

American market. Roughly ninety thousand patent applications are made each year in the United States.

In this section, the elements of patenting in the United States will be summarized. While patent systems vary considerably from country to country, the patent application process described here is fairly typical. The U.S. Patent and Trademark Office lists six basic steps for obtaining and maintaining an American patent:[3]

Step 1. Make Certain the Invention is Practical. The fact that an inventor has obtained a patent on an invention is no assurance that the invention will be commercially viable. Fewer than half of patented inventions are actually commercialized. The Patent and Trademark Office recommends that inventors attempt to determine the practicality and commercial merits of their inventions prior to seeking patent protection. The inventor may determine that there is no sense in incurring possibly substantial costs in time and money to patent something with no commercial value.

Step 2. Have Witnesses and Maintain Good Records. In the United States, unlike most other countries, disputes involving simultaneous discoveries are resolved in favor of the inventor who can prove that he was the first to conceive and develop an invention. Thus, it is important that an inventor keep very good records of his work and that his achievements are confirmed by witnesses. When what are called interference proceedings are held by the Patent and Trademark Office to determine which of two or more inventors has priority over a given invention, the documentation and witness testimony will be significant in determining priority.

Step 3. Perform a Preliminary Patent Search. An inventor cannot obtain a valid patent if his invention is anticipated by an earlier printed publication or patent in any country, or by commercial use in the United States. Thus, it is important that he search the patent files to examine the prior art. A patent search serves three useful functions. First, the search may alert the inventor to the fact that his idea or invention has already been patented. If this is so, then the inventor need not assume the expense of filing for a patent. Second, even if no prior art can be found that exactly duplicates an inventor's invention, he may learn through a patent search that existing technology is superior to what he has developed. This knowledge may lead him to abandon the attempt to obtain a patent, since the existence of superior technology may not allow his invention to be commercially viable. Third, by examining inventions similar to his, the inventor may develop an idea of how best to patent his own invention so that it is afforded the strongest protection possible.

Patent searches are generally conducted by professionals who are registered with the Patent and Trademark Office. The searches are usually undertaken in the Patent and Trademark Office search room in Arlington, Virginia, a suburb of Washington, D.C. This room contains the largest collection of U.S. patents in the world.

Step 4. Study Patents Found in the Search. The patent searcher will present the inventor with copies of patents that have a bearing upon his invention. The inventor should study these patents carefully in order to determine the extent to which his invention has new features that are different from the prior art. No patent claims can cover old features. Any patent that an inventor obtains will cover features that make his invention different from those described in prior patents.

Step 5. Prepare the Patent Application. Patent applications should be prepared by attorneys or agents registered with the Patent and Trademark Office. It is very important that the application be prepared skillfully. Two parts of the application are particularly important: the patent specification and the claims.

The law requires that the patent specification describe the invention in such a way that a person who is skilled in the field of invention can make and use it. It may be necessary to include drawings of the invention if they clarify the description.

The claims are unquestionably the single most important part of the application. They must describe the unique features of the invention and distinguish it from previous inventions. Writing the claims entails a delicate balancing act. On the one hand, they should be broad enough to cover as much ground as possible, since claims define the boundaries of the inventor's patent rights. On the other hand, they should not be so broad and vague that they will be indefensible in a patent infringement case. Poorly written claims are an invitation to patent infringement.

Step 6. Submit Application. Each application that is submitted to the Patent and Trademark Office is assigned to a patent examiner. The examiner reviews the application and then searches the patent and other literatures to locate inventions similar to the one covered in the application. He examines the claims carefully in respect to the prior art. He may reject some or all of the claims on the grounds that they are not sufficiently different from the prior art, or that they are written too vaguely, or that they reflect new features that are obvious to individuals having skills in the field of invention. Rejection of some claims by the patent examiner is a common occurrence. The examiner notifies the patent attorney or agent of the rejection of claims and allots him a specified period of time to amend the application.

In preparing the amendment, the patent attorney or agent and the inventor review the patents and literature cited by the examiner. After the review, they decide whether to abandon the patent or to continue prosecution. If they proceed forward, they modify the application to meet the examiners objections and to spell out reasons for believing the patent should be granted. The amendment is then submitted to the Patent and Trademark Office.

The patent examiner reviews the amendment and may reject some of the claims, all of the claims, or may determine that the patent should be granted. The amendment procedure continues until the examiner grants a patent, or declares that rejection is final.

The prosecution of a patent entails substantial literature searches and communications among the patent examiner, the attorney or agent, and the inventor. Not surprisingly, prosecution of a patent can take a long time. Until the mid-1970s, it would typically take three years for patents to be granted. However, after the Patent and Trademark Office expedited the review procedure, examination time was reduced to about a year and a half to two years.

International Patenting

An American inventor who has developed a new product that is ripe for commercialization in the American market most certainly will give serious consideration to patenting it in the United States. Should he obtain a patent in the United States, he has monopoly rights over his invention for seventeen years from the date the patent is issued. However, his U.S. patent will not protect his product from infringement in foreign markets. For example, if he has no patent protection in West Germany, there is no legal restriction keeping West German firms from manufacturing and marketing his product in that country or in any other where the invention is not patented. It should be noted, though, that West German firms cannot sell the product in the United States and cannot manufacture it in the United States for sale overseas without the approval of the inventor.

It is evident that, if an inventor wishes to sell or license his wares overseas, he must consider patenting his inventions in foreign countries as well as in the United States. Unfortunately, obtaining patents in more than one country can be a complex and costly undertaking. It is not simply a matter of taking the U.S. patent and filing it, in that present form, in other countries. Each country has its own patenting system and no two systems are alike. The basic objective of any patent system is the same throughout the world: to encourage the introduction of inventions into the economy by conferring upon the inventor monopoly rights over his invention. However, a great number of different means are employed to achieve this objective. Some of the distinguishing features of the different patenting systems will be discussed in the next section of this chapter.

It has long been recognized that the existence of a multitude of national patent systems, each with its own rules, makes it enormously difficult for an inventor to obtain patent protection in more than one country. Consequently, the Paris Convention for the Protection of Industrial Property was put into effect in 1883 to inject some order into international patenting. The Paris convention, which today has some ninety signatories, has been revised a number of times since its inception. It provides that each member country extend to the citizens of other member countries the same rights in patent and trademark matters that it extends to its own citizens. Furthermore, it allows someone applying for a patent in one of the member countries up to twelve months after the first filing to file for a patent in other member countries. The effective filing date for all applications is that of the first filing. Establishing the earliest filing date possible may be important in determining an invention's priority over competing inventions and in deciding who is the original patent owner.

In general, the single most important issue determining whether an inventor will file for patent protection overseas is the cost of filing and maintaining patents in other countries. If the costs were insubstantial, then there would be no real problem; the inventor would file for patent protection in as many countries as possible, just to insure as broad protection of his invention as possible. However, the cost of filing and maintaining patents overseas can be substantial, easily ranging in the thousands of dollars per patent. There are several elements leading to these high costs. First, the inventor must pay the standard filing fees, which typically cost several hundred dollars. Second, he must tailor the application to meet the specific requirements of the country in which he wishes to patent his invention. In doing so, he incurs administrative costs. Third, if the target country requires demonstration of the novelty of the invention, there will be costs in time and money, generally including the hiring of local personnel to see to the details of the patent examination. Fourth, if the patent is being filed in a non-English speaking or writing country, translations must be made and this again adds to the cost of filing. Finally, once a patent is granted, a substantial number of countries require the payment of annual renewal and maintenance fees to keep the patent active. Frequently, these fees escalate from year to year. The purpose of escalating fees is to cause holders to let lapse patents that they cannot or do not intend to commercialize.

Given the high cost of obtaining patents overseas, one must calculate the anticipated benefits of such an action to see if they outweigh the cost. If they do, then one has good reason to pursue patent protection overseas. What then are the benefits? Clearly, the most significant benefit is the same as for domestic patenting. That is, obtaining a patent overseas gives an inventor monopoly rights over his invention in other countries. With the conferral of these rights, no one can product or sell his product without his expressed consent in the countries where he holds a patent. The inventor can produce the patented product himself in these countries, or else he can license others to do so.

Another benefit to the inventor is that he may be able to use a particularly valuable patent as a bargaining item to extract desired concessions from institutions in a foreign country. A good example of this is the case of IBM in its dealings in Japan in the early 1960s. In 1957 the Japanese Diet (parliament) passed the Provisional Electronic Promotion Laws. Among other things, these laws prohibited the establishment of foreign subsidiaries in Japan in the computer industry. Soon after the passage of these laws, it became apparent that Japanese industry needed key technology patented by IBM in Japan. In order to obtain access to this technology, the government allowed IBM to establish a wholly owned subsidiary in Japan in exchange for release of the key technology.[4]

Still another benefit of possessing patent protection overseas is that it may enable the inventor to encourage foreign participation in a joint venture. If a foreign firm sees that the only way it can produce an attractive product whose patent is owned by an outsider is through collaboration, that firm will have substantial inducement to participate in a joint venture. Or else, parties may choose to swap patents in order to obtain access to each other's technology.

Once an inventor has determined in principle that he wants to take out a patent on his invention overseas, he must decide which country to focus on. Many different factors must be taken into account in making this decision. First, he will have to take into account a number of procedural matters concerning patenting in different countries. For example, he will not attempt to acquire a patent where he is disqualified from patenting. This may occur if he has published information on his invention or begun commercial operations on it prior to filing for a patent. In the United States, prior publication or commercialization is permitted, so long as it occurs less than one year before the patent application. In a number of other countries—for example, France and Italy—prior publication or commercialization is a bar to patenting. There are several other procedural considerations that the inventor will want to take into account, and these will be discussed in the next section of this chapter.

Second, he will want to give serious consideration to patenting his invention in those countries where he will manufacture and/or market it. Clearly, it is in such countries that it is most important that the invention be protected.

Third, the inventor should consider patenting his invention in countries where it is likely that a local firm will license the technology. In this case, the inventor would not directly manufacture or market the patented product, but he nonetheless can earn a return on it through royalty payments made by the licensee.

In general, there is little reason to go through the effort of patenting a product in places where there is little or no demand for it or insufficient capability to produce it. Major firms that engage in large-scale international patenting of their inventions typically restrict their patenting efforts to a small number of key countries with high technological capabilities and strong internal markets, countries such as Japan, France, and West Germany. Less-developed countries often lack adequate production capabilities and internal markets for high-technology items, so these countries do not generally attract much patenting interest. (A detailed consideration of patenting in LDCs is given in chapter 10.)

Variations in Patent Systems

Patent systems vary substantially from country to country. There is no single global patenting system to which all countries subscribe. In recent years steps have been taken to introduce an element of uniformity to international patenting. The Patent Cooperation Treaty, for example, came into effect in 1978. This treaty permits the filing of a single patent application in a uniform format that applies to designated countries. While the application entails a single international patent search, it is still examined separately by the different countries. (The success of this approach is undetermined at present. It will have to be used for a number of years before its effectiveness and acceptability are determined.)

Despite attempts to make international patenting more uniform, the system remains fragmented. A number of the more important features distinguishing different national patent systems are:

Patent Life Times. As previously stated, patent lifetimes vary from country to country. Not only do the lifetimes vary, but the date from which the start of the patent life is calculated differs among countries. For example, in the United States, the seventeen-year life of the patent commences on the date the patent is granted, while in many other countries the clock begins ticking on the date the inventor first applies for a patent.

An inventor should estimate the effective life of a patent in different countries and should take this estimate into account when deciding whether or not to apply for a patent in any of these countries. Circumstances can considerably reduce effective patent lifetimes. For example, even though the life of a patent in the United States is seventeen years from the time of the granting of the patent, if the patent is for a new drug, it may not be marketable for several years because of laws governing the testing of drugs. By the time the Federal Drug Administration approves a drug for marketing, only a dozen or so—not seventeen—years may remain before the patent expires.

Prior Description or Use of Invention. Some countries are liberal in allowing an inventor to describe work or commercialize it prior to applying for a patent. These countries have liberal standards regarding prior use, since they do not require that the invention receive its first public exposure through the patent application. Other countries have strict standards regarding prior use. In these countries, prior description or commercialization of an invention disqualifies it from receiving a patent.

When considering standards of prior use from a worldwide perspective, we find that countries generally fall into one of three categories. First, there are countries that allow an invention to be described or commercialized for a specific period prior to making a patent application. The United States, for example, has a one-year time limit. Countries falling into this category, in addition to the United States, include Canada, West Germany, Japan, and Switzerland.

Second, some countries allow no prior description or commercialization of the invention at all. Included here are Mexico, France, Italy, and the Scandinavian countries.

Third, a number of countries have no time limit on prior description or commercialization of the invention, provided these activities occur outside the country. Included here are Israel, Great Britain, Australia, and New Zealand.

It is obvious that anyone wishing to patent an invention overseas will have to determine the standards of prior use in the target countries, particularly if he has already described his invention in, for example, an article, or has begun commercialization of it. Prior publication or commercialization will preclude patent protection in countries with strict standards. Whether such acts prevent protection in other countries depends upon their policies concerning the allowable time limits for the early description or commercialization of the invention.

Obviousness of the Technology over Prior Art. Countries vary considerably over the degree to which an invention must be different from the prior art. For example, in West Germany, Japan, and the Scandinavian countries, the new invention must be substantially different from the prior art. The new invention cannot be a simple modification of existing technology. In contrast, Belgium has no such requirements. Patents are conferred automatically in a very short time period (within a month of the application) and no searches are undertaken to determine whether the invention is different from existing technology. In the United States, Great Britain, and Canada, there is a moderate requirement for obviousness of an invention: within certain bounds, a new invention can be derived from and similar to existing technology.

Publication of Patents. In a number of countries, including the United States, a description of an invention is published only after a patent is conferred. In other countries, patents are published in a specified time period following the filing of the patent, typically in eighteen months. West Germany, the Netherlands, Japan, and France follow this system.

When the description of an invention is published, it alerts competitors to the activities of the inventor. Furthermore, the description allows competitors to determine how to produce the invention. Consequently, the publication of an invention's description is a significant event for the inventor. West Germany, for example, is a country that requires publication of the patent application eighteen months after filing. Furthermore, as previously noted, West Germany is strict in requiring inventions to possess novelty over prior art. Thus, it is possible in a country such as West Germany for an inventor to file for a patent, have a description of his invention published, and then be denied a patent. In such a case the invention is publicly disclosed and yet the inventor does not obtain patent protection. In retrospect, the inventor would have been better off not to have applied for that particular patent in West Germany.

Interference Proceedings. When two or more inventors file for patents on the same invention at roughly the same time, who should be awarded the patent? In most countries, the inventor with the earliest application is awarded the patent. However, it is clear that the date of a patent application does not necessarily reflect the true priority of an invention. The second applicant may have been cautious about the merits of his invention and tested his invention extensively before applying for patent protection, even though he actually produced his invention earlier than the first applicant; however, in countries where patents are awarded according to application date, he would be denied a patent in favor of the first applicant.

In a country such as the United States, patent awards are made on the basis of the date of the actual discovery, not on the basis of the application date. When two or more inventors file for patents on the same invention, interference proceedings are held in order to determine who first conceived and produced his invention. In such a system, an inventor is strongly advised to carefully document the steps taken in coming up with his invention, since such documentation may enable him to claim priority.

Opposition Proceedings. In some countries, such as West Germany and Great Britain, inventions that are viewed as worthy of patenting are published for a limited time prior to conferring the patent. The public is invited to examine the description of the invention and to oppose the granting of a patent if, for example, it can be demonstrated that the invention is insufficiently novel. While these proceedings consume valuable time and re-sources, they help assure that patented inventions are unique and worthy. In countries with opposition proceedings, the courts are more apt to uphold the validity of a patent in an infringement case than in countries without them.

Working Requirements. Some countries require that a patent be commer-cialized within a specified time. If it is not, then the patent holder may be re-quired to license the technology to someone willing to commercialize it. If the technology still is not commercialized, then the patent may be revoked. While the Paris Convention allows the establishment of working require-ments for a patent, it places a number of restrictions on enforcement, so that, in practice, many years can pass before a patent holder is actually re-quired to license his technology. Countries with working requirements have them in order to clear the books of trivial inventions and to reduce the in-cidence of defensive patenting.

Licensing

As explained in chapter 5, licensing is the process whereby the owner of a technology allows others to use or develop it under controlled circumstances, in return for remuneration, frequently in the form of royalties.

Licensing has become an increasingly attractive technology-transfer in-strument in recent years. No doubt, in their efforts to acquire technology through licenses, some prospective recipients have an eye on the very effec-tive use of licenses by the Japanese. For their part, potential technology donors often view licensing as a low effort and lucrative way to earn returns on their R&D undertakings. In the next several pages, some of the benefits and shortcomings of licensing will be discussed.

Benefits to the Licensee

Whether or not a firm has strong internal R&D capabilities, it should con-sider licensing technology from others. Some of the ways that a licensee may benefit from entering into such agreement include:

Savings in R&D. When a company licenses technology from others, it can save considerably on R&D expenses. By purchasing technology through a license, the company is buying the fruits of someone else's R&D. A major implication of this is that by pursuing an active licensing policy, the licensee will probably not be first to market. Rather, he will be pursuing a follow-the-leader strategy. If he is not first to market, it is unlikely that he will gain the market share he might have had he been, as the saying goes, the firstest with the mostest. If the licensee hopes to be competitive and gain a significant market share, he will have to produce the good more cheaply, or modify it to make it more attractive, than competing goods.

Avoidance of Duplication of Effort. A firm with good R&D capabilities may wish to develop a new product only to discover that another firm has already developed the technology to produce it. The first firm faces two choices: it can incur substantial R&D expenses to develop a new way to produce the product, being careful not to infringe on the second firm's patents; or it can attempt to license the technology from the second firm and save on R&D expenses. If it seriously investigates this second possibility, it may find that the savings are so substantial that it would be foolish to attempt to duplicate the efforts of the second firm.

Avoidance of Patent Infringement. Occasionally two firms independently develop the same technology and risk patent infringement difficulties. A full-blown patent infringement suit could cost each party from fifty thousand to several hundred thousand dollars in legal expenses. Furthermore, the outcome of the suit would not be certain, and it could well be that one party would lose a good deal of control over the technology he had labored to develop. One way to resolve the conflict amicably would be for one party (or both) to license the other's technology, thereby obtaining explicit permission to exploit the technology and removing the threat of a patent infringement suit.

Picking up Technological Skills. Sometimes a firm lacks the skills necessary to produce a good and will license a technology in order to learn how to do it. In this case, licensing serves as an instructional tool, enabling the licensee to learn by doing. Of course, for this approach to work, the licensing agreement would have to stipulate that whatever is necessary to give the technology recipient the needed skills would be provided by the donor. Lack of technological skills is a common occurrence in LDCs, so this particular aspect of licensing is attractive to them.

Developing New, and Bolstering Existing, Product Lines. Some firms, such as the 3M Company, are loathe to go outside their organization for advice and material assistance. Others, such as Du Pont, eagerly look outside their organization for anything that can help corporate performance. This second type of firm will license technology from outside when it contributes to a corporate strategy to develop new product lines or flesh out existing lines.

Licensing May Give a Firm Its Only Access to a Superior Technology. An inventor may develop a vastly superior new technology that is well protected by a patent or trade secret. In such a situation, one's only access to the technology may be through a licensing arrangement with the inventor (assuming, of course, that he is willing to offer others access to his technology). A well-known example of this is the development by Pilkington Brothers of the float-glass process for producing sheet glass. The new process produced sheet glass far more cheaply than the old, which required the mechanical grinding of sheet glass to rid it of surface imperfections. Pilkington Brothers, a British firm, decided that it would license its technology to the other sheet glass makers. Because of the obvious cost effectiveness of the Pilkington float-glass process, glass makers around the world eagerly entered into licensing arrangements.

Benefits to the Licensor

There are many tangible benefits to a firm licensing its technology. Of course, before it actually licenses its technology, the firm must determine whether the benefits outweigh the cost of losing some control over it. The major benefits to the technology licensor include:

Earning Returns on R&D Efforts through Royalties. If a firm possesses a unique, attractive technology that is protected by a patent or is a well-kept trade secret, the firm may jealously guard its monopoly power over the technology and be reluctant to share it with others through a license. This is particularly true for brand-new technologies that give the firm's products a sharp edge in the market place. However, circumstances may exist that would lead a firm to license its technology and earn royalties on it.

First, a firm can profitably license technologies that it has developed in its laboratories but has decided not to exploit itself. A firm may choose not to exploit its own technology, for example, when a discovery results in a product that is not in the firm's line of business, or when the firm decides that it has insufficient resources to develop and market it.

Second, a firm can profitably license to others technologies that are outdated for its operations. When a new technology replaces the old, the firm can try to earn royalties off of it rather than relegating it to the rubbish heap, where it would earn nothing. The technology may be obsolete to the firm, but may still have substantial value to others.

Third, a firm may find that intense competition in the market depresses prices on its products and does not allow its technology to earn adequate returns. Having concluded this, the firm may decide to increase the returns on the technology by licensing it to others.

Earning Returns on Goods and Services Supplied the Licensee. Significant returns can be made not only through royalties, but by means of sales of goods and services to the licensee. Particularly at the outset of an agreement, a licensee is often heavily dependent upon the licensor for know-how, equipment, spare parts, managerial assistance, and so on. The licensee obtains these things at a price from the licensor. If the licensor is the sole supplier to the licensee of certain raw and manufactured materials, the state of dependence can continue indefinitely, and the licensor can have a steady source of income for a long time. It should be noted that many countries have laws forbidding tied licensing agreements, where the licensee is forced to purchase supplies from the licensor. However, when the licensor is the sole possessor of needed supplies, the licensee has little choice but to deal with it.

Establishing Valuable Contact with Other Companies. A licensing arrangement may enable the licensor to establish a valuable working relationship with other companies and may make possible highly profitable joint ventures. Very often, an agreement requires that the licensee and licensor work closely together. In working closely, they may each develop an appreciation of the strengths of the other. Ultimately, they may find that their strengths dovetail nicely, and this may lead to the establishment of a mutually beneficial joint venture.

Penetrating New Markets. A licensing agreement may enable a licensor to penetrate new markets. This can be seen in the case of an agreement between International Harvester and the Polish government to produce, under license, bulldozers, loaders, and pipelayers. Prior to this agreement, International Harvester had no contact with Eastern European countries. The company figured that with the Polish manufacture of International Harvester equipment, Eastern Europeans would become familiar with company products. The licensing agreement was thus a first step taken by International Harvester to penetrate the Eastern European market.[5]

Reducing Incentives for Others to Produce Competing Technologies. Sometimes by allowing others to share its technology through licenses, a firm can make it less attractive for others to develop competing technologies. The earlier mentioned Pilkington float-glass process is a case in point. After Pilkington Brothers developed it, they had a choice either to be the sole users of the process or to license it to competing sheet glass manufacturing companies. Pilkington opted for the second course of action, partially in order to keep competitors from developing improved methods for the production of sheet glass.[6] So long as these competitors could use the Pilkington process on reasonable terms, it was not cost effective for them to explore alternative means for producing sheet glass.

Why Not License? Licensee's Perspective

We have just seen that there are many benefits associated with licensing. There are a number of disadvantages as well. From the point of view of the licensee, three disadvantages stand out:

The Licensee is Heavily Dependent upon the Licensor. The dependence of the licensee on the licensor basically has two dimensions to it. First, the license agreement establishes conditions that restrict the actions of the licensee. For example, it may limit the licensee to a relatively small market territory; or it may require strong and costly quality-control procedures; or it may force the licensee to buy supplies from the licensor; and so on.

Second, the licensee is often dependent on the technical expertise of the licensor, particularly at the outset of the agreement, or if the licensee lacks adequate technical capabilities. If the licensor withholds this expertise, the licensee could face serious trouble.

The Licensee Lacks Initiative. A licensing strategy is to a large extent a reactive strategy. The licensee can only acquire technologies that others make available. Consequently, the licensee often does not possess the advantages of initiating new products. Rather, his advantage must be based on other factors. For example, low production costs, innovative marketing strategies, reputation for high-quality production.

Licensing May Result in Underdeveloped R&D Capabilities. Firms that heavily license other firms' technology may determine that they have little reason to develop strong R&D capabilities themselves. As a result, their R&D capabilities may remain (or be) atrophied. So long as this situation prevails, they will be dependent upon outsiders in a very critical area. However, a firm that licenses technology from others need not lose its R&D capabilities. Du Point has licensed outside technology effectively and still maintains strong R&D capabilities. Nonetheless, when licensing is the cornerstone of a firm's technology strategy, R&D may very well suffer.

These observations appear to apply to nations as well as firms. Japan is noted for its borrowing of technology from other countries. No one can doubt that this policy has been very profitable for the Japanese. However, it is also generally agreed that outside of a few well-publicized areas (for example, electronics and computers), Japanese research performance is often run-of-the-mill, particularly in the more basic research areas. To a certain extent, Japan's modest research achievements are probably linked to its heavy emphasis on licensing and modifying the research accomplishments of others. The Japanese government seems to be aware of this, and in recent years has attempted to dramatically upgrade the national research undertaking.

Why Not License? Licensor's Perspective

From the licensor's perspective, there are at least two disadvantages to licensing his technology, one major, the other relatively inconsequential.

Technology give-aways. The single most pronounced danger in licensing technology is that a firm may be giving away hard-earned technology. Ultimately, this technology may strengthen the position of competitors, so that despite short-term earnings of royalties, the firm's position may be eroded in the long run. This also holds true at the national level. Many American economists have expressed the fear that the liberality of U.S. firms in licensing their technologies to firms in other countries has contributed to the lack of U.S. competitiveness in international markets, where erstwhile foreign licensees are now in head-on competition with American companies.

Administrative burden of licensing. A second disadvantage of licensing to the licensor is the administrative burden it imposes on him. Licensing agreements require constant monitoring. The licensor must be responsive to requests by the licensee. If the agreement allows the licensee to employ the licensor's trademark, then the licensor must be sure that the licensee is producing quality products, so that the trademark is not damaged. The licensor may have to establish a licensing department to handle the added administrative burden.

While these administrative burdens are a disadvantage, they will be more than offset by earnings generated in a lucrative licensing effort. If licensing is handled as another product by the firm, rather than as an interesting aberration, then the expenses of administering agreements are conceptually no different than other routine business expenses.

The Licensing Agreement

There are few, if any, hard-and-fast rules for licensing. Licensing agreements are negotiated between two or more parties. As in the case of any negotiation, the outcome is a product of the needs of the parties and their negotiating skills. Consequently, licenses can be unique, custom-made documents. The terms of the license, provisions for administering it, forms of remuneration, and the actual wording of the document are limited only by the needs and imaginations of the licensor and licensee. Nonetheless, when we deal with licenses at a fairly high level of abstraction, we can identify typical features associated with them, some of which will be discussed here.

The typical license describes the rights and responsibilities of the licensor and licensee in a straight-forward fashion. It specifies how the licensor will transfer technology to the licensee; limitations on the use of the technology by the licensee; and mechanisms by which the licensor will be remunerated for use of the technology.

How the Technology Is Transferred to the Licensee. This will depend largely on the nature of the technology. The transfer may entail, when appropriate, the provision of blueprints and drawings, engineering data, information on production techniques, and a statement allowing the licensee to share the licensor's patent rights. Arrangements may also provide for training by the licensor of the licensee's personnel. Training may occur at the licensor's facilities as well as on-site. Finally, provisions are often included requiring the licensor to update the licensed technology.

Restrictions. There are a wide range of limitations that a licensor can place on the licensee's use of his technology. The legality of some of these restrictions varies from country to country. Third World countries in particular feel that many of these restrictions have a deleterious effect on technology-transfer and they have been trying to curtail them by creating a code of conduct for technology transfer (see chapter 10). For their part, licensors argue that many of these restrictions are valid mechanisms for protecting their interests, and without them, no licenses would be issued. Among the more prominent restrictions are:

Tie-in Clauses. These clauses require the licensee to obtain certain materials and/or spare parts only from the licensor.

Package Licensing. In order to obtain a license for needed technology, the licensee is required to license additional technology that it may not want.

Grant-back Provisions. These provisions require that any improvements the licensee makes to licensed technology revert back to the licensor.

Quality Control Clauses. Here the licensor requires that licensed products manufactured by the licensee meet certain quality standards.

Field of Use Restrictions. These restrictions give the licensor the right to control the fields in which the licensed technology can be applied.

Production Volume Restraints. These are requirements restricting the quantity of production by the licensee.

Export Restrictions. These restrictions allow the licensor to determine where the licensee can sell his products.

Tie-Out Clauses. These clauses forbid the licensee to obtain complementary or competing technology from sources other than the licensor.

Price Fixing. Here the licensor attempts to dictate the selling price of licensed products.

Management Participation by Licensor. Occasionally, the licensor wishes to participate in the management of the licensee enterprise in order to assure that the licensed technology is exploited properly.

Restrictions on Use After Expiration of Agreement. These restrictions are an attempt by the licensor to assure that the licensee will not continue to utilize the licensed technology once the licensing agreement lapses.

Remuneration. The forms of remuneration through which a licensor receives payment from a licensee are numerous. The most common form is a royalty payment based on the volume of sales of the licensed product. Sales are generally viewed as the best basis for calculating royalty payments because sales levels tend to be fairly stable over time. In contrast, profits are highly volatile and subject to accounting manipulations. If the royalty payments were based on profits, when the licensee suffered losses over a given period, the licensor would receive no royalties. However, with royalties based on sales, the licensor would receive payments, even if the licensee's operation was not profitable.

The actual royalty rate will, of course, depend upon the nature of the licensed product and the bargaining abilities of the licensor and licensee. Typically, royalties range from 1 to 5 percent of sales.

A second form of remuneration is the downpayment fee. Such a fee is paid at the outset of the agreement. It serves several functions. First, it demonstrates a commitment on the part of the licensee to produce and market the licensed product effectively. It is to the advantage of both parties that the licensee be successful in producing and selling the good. The downpayment will help assure the licensor that the licensee is serious about the undertaking.

Second, a downpayment may guarantee that the licensor will recover costs associated with administering the license, for example, start-up costs, consulting fees, costs of blueprints and drawings, and training costs.

Third, the downpayment may be a premium that the licensor charges to cover the special risks and expenses associated with doing business in the target country (for example, restrictive labor laws, frequent strikes, political unrest).

A third form of remuneration is the acquisition of equity interest by the licensor in the licensee's organization. In this instance, the licensor may be more concerned with long-run gains resulting from partial ownership of the licensee firm than from more immediate cash gains coming from royalty payments.

A fourth form of remuneration is covered in a buy-back provision that may be included in the license document. With a buy-back provision, the licensor is allowed to buy back some of the output of the licensee at a discount price. The buy-back provision can be appealing to both parties. By buying back goods produced by the licensee, the licensor may acquire goods that it needs for sales at below its own production costs. Furthermore, it can do this without incurring the expenses of expanding plant capacity. To the licensee, the buy-back provision may be appealing because it guarantees that a portion of his output is purchased.

Cross-licensing and Pooling

With cross-licensing, two firms license technology to each other in what amounts to a swap. In this way, each firm is able to obtain valuable technology with little or not cash outlay. If two firms are in the same industry and wish to undertake a cross-licensing agreement, they must be careful not to violate the antitrust regulations that apply in the countries in which they operate.

With pooling, a number of firms place their proprietary technology into a common pool. If other firms are not given access to the technology in the pool, then the pool may be in violation of local antitrust laws.

Joint Ventures

The term joint venture is used in a number of different ways. In the broadest sense, it can suggest any kind of collaboration between two or more business entities. However, it also has a more precise meaning that is commonly used. That is, a joint venture entails the establishment by two (or more) business entities of a third entity which serves a specific function. Establishment of joint ventures achieved a degree of popularity in the 1970s, although by the 1980s they lost some of their allure after examinations of their performances showed mixed results.

Motivations for Joint Ventures

There are many reasons why two companies might wish to undertake a joint venture. The following list highlights the most salient reasons, but it is by no means exhaustive.

Accommodation of Foreign Investment Laws. A number of countries do not allow the operation of wholly owned subsidiaries on their soil. In some cases, while they may allow equity interest in a local operation by foreigners, this interest must be less than 50 percent. In a situation such as this, a joint venture with a local firm may be a necessity for a foreign firm desiring to make an investment in the country.

Resources Sharing. A joint venture may be appealing to a firm wishing to employ the resources of another firm. The resources may be financial, in which case the firm undertakes the joint venture for cost-sharing purposes; or they may be material, as when a firm making high-grade ball bearings enters into a joint venture with a specialty steel supplier.

Skills Sharing. One specific form of resource sharing that is often a motivation for the creation of joint ventures is skills sharing. In this case, two firms with complementary abilities get together in a joint venture in order to build on their respective strengths. For example, one company might have strong R&D and marketing capabilities, while the other might have strong production capabilities. In the joint venture, the first company would be charged with undertaking R&D and marketing tasks, while the second would engage in production.

Use of Expertise of Locals. A foreign firm may enter a new country with great trepidation, realizing that it has little knowledge of local markets and regulations. In this case, it may wish to enter into a joint venture with a local firm in order to capitalize on its knowledgeability regarding local markets and regulations.

Risk Reduction. There are many different ways in which a joint venture can reduce the risks of the partners. Three examples will be offered. First, if the joint venture is structured as a corporation, loss will be limited to the assets of the joint venture and the partners will be protected from creditors. Second, in a country that is antipathetic to foreign investments, risk of nationalization may be reduced if an outsider sets up a joint venture with a local firm. Third, in a very large-scale project (for example, off-shore drilling for oil), setting up many smaller joint ventures may lessen risk, since it is spread over many smaller projects.

First Step in a Merger/Acquisition. A joint venture provides firms with a good way to get to know each other well, since through it they are able to observe closely how they each function. Consequently, if one firm wishes to acquire another, or if two firms desire to merge, they may wish first to enter into a joint venture—a trial marriage, of sorts. If they work well together, a merger/acquisition might be encouraged.

Heavy Debt Financing. Corporations are often careful to maintain a solid debt/equity ratio. However, if a firm operates through a joint venture, it has the option of undertaking heavy debt financing for a project without having the debt show up on its balance sheet.

Structuring the Joint Venture

Joint ventures can be structured in many different ways. The primary consideration in structuring a joint venture is how ownership and control are distributed among the partners. Control and ownership often, but not always, go together. That is, the majority owner usually is the controlling partner.

In dealing with the structuring of ownership, attention generally focuses on two possibilities: first, each party has 50 percent interest in the joint venture; and second, one party clearly has majority interest while the other clearly has minority interest. Each of these approaches has certain implications associated with it. For the purposes of our discussion, we will assume for a moment that ownership and control are one and the same.

In the case of fifty-fifty ownership, each party is clearly reluctant to give up controlling interest in the joint venture. One problem that may arise in this situation is the possibility of deadlocks in decision making. If it is likely that deadlocks will arise, the partners in the joint venture should work out some mechanism whereby decisions can be made. For example, the joint venture could be structured so that each partner has 49 percent interest, with 2 percent interest given to a third party, who would serve as tie-breaker in the eventuality of deadlocks. Or else, the partners can establish a protocol for resolving deadlocks. For example, decisions can be made by means of a management committee with equal representation by each partner, with one or more individuals added to the committee to break ties.

Where one partner is clearly the majority owner, he most frequently takes the lead in decision making, especially when the minority partner is not interested in playing an active management role. The importance the minority partner attaches to his participation in management is generally tied to his role in the joint venture. For example, if his role is simply to provide necessary supplies to the majority partner, it is not likely that he will be very active in the management of the joint venture. In general, the more significant the role of the minority partner, the greater his interest in having a material impact on decision making.

There are a number of ways that ownership and control can be kept separate. Consider, for example, a company that is forced to have minority interest in a joint venture because of foreign investment laws. If the company is the driving force behind the venture—supplying marketing expertise, know-how, management skills—it will want to have substantial control over developments in the joint venture. One way to do this is to require that the key management in the joint venture be chosen by the company. This will help assure that day-to-day management decisions are to the liking of the minority partner. In addition, the joint venture agreement may require that important decisions be made with a super majority. That is, if the minority partner has 40 percent interest in the joint venture, 61 percent voting is necessary to make major decisions (for example, to restructure the joint venture).

Another way for a minority partner to have control out of proportion to its share of the joint venture is to set up a nexus of interlocking joint ventures with different firms. These can be set up in such a way that, although this company does not have majority interest in any one joint venture, the cumulative effect of its position in all the interlocking joint ventures allows it to have more clout overall than any of its partners.[7]

In the final analysis, if the joint venture is to be successful, the partners will have to cooperate with each other, irrespective of their ownership position. Chances are that the reason they entered into a joint venture in the first place was because each needed either the resources or skills of the other. That is, there is a mutual dependence of the partners upon each other. If the partner with 90 percent interest in the venture constantly thwarts the desires of the 10 percent partner by virtue of its majority holdings in the venture, the prospects for the joint undertaking are not very bright. Cooperation—not confrontation—is a necessary condition for the smooth functioning of a joint venture.

Problems with Joint Ventures

Joint ventures can be fraught with problems. Many of these problems are rooted in the fact that a partner in a joint venture is essentially an independent entity over whom only minimal control can be exercised. For example, if one partner in a joint venture unilaterally determines that capital shortages he faces in his other operations prohibit him from making capital contributions to the venture, the other partner may find himself in a real bind through no fault of his own

In order to control these kinds of problems, joint venture agreements are often filled with legalisms that try to anticipate any and all problems. For example, in order to guard against a joint venture being abandoned midstream, the partners may include in the agreement provisions guaranteeing that the project be completed, requiring the purchase of a certain quantity of the manufactured product at a specified price, and requiring that funds be set aside by each partner to cover operating losses that may be incurred. Coping with these legalisms, in addition to meeting sundry bureaucratic requirements of each partner, often means that joint ventures have higher administrative costs than regular businesses.

A firm that is considering entering into a joint venture for the purpose of acquiring technology should certainly weigh the advantages and disadvantages of a joint venture against those of a licensing agreement. Many factors will go into making a decision on which mechanism to choose. One of the most significant is the technological absorptive capacity of the technology recipient. If his absorptive capacity is very low but his partner has strong technological capabilities, he may prefer a joint venture.

In this case, the recipient's technological deficiencies will not negatively affect the venture. On the other hand, if the technology recipient has a superlative absorptive capacity, he may wish to license the needed technology from a donor and go it alone in his venture. By traveling this path, he will not be restricted by the actions or desires of a partner (although he will still be restricted in his actions by the terms of the licensing agreement).

Notes

1. Patent and Trademark Office, *Patents & Inventions: An Information Aid for Inventors* (Washington, D.C., 1978), p. 1.

2. T.M. Noone, "Trade Secret vs Patent Protection," *Research Management* 21 (May 1978):22.

3. Patent and Trademark Office, *Patents & Inventions*, pp. 2-11.

4. J. Baranson, *Technology and the Multinationals* (Lexington, Mass.: Lexington Books, D.C. Heath and Company, 1978), p. 74.

5. E.W. Hayden, "Transferring Technology to the Soviet Bloc: U.S. Corporate Experience," *Research Management* 19 (September 1976):17-23.

6. B.C. Twiss, *Managing Technological Innovation*, 2nd ed. (New York: Longman, 1980), p. 59.

7. S. Gullander, "Joint Ventures and Corporate Strategy," *Columbia Journal of World Business*, 11 (Spring 1976):110.

8 National Science Policies

Introduction

Today governments play such a central role in developing and maintaining national scientific and technological capabilities that it seems as if it always must have been so. In point of fact, governments have only recently been actively involved in shaping science and technology in their countries. There was a time, not so long ago, when governments played a minimal role. Science was largely in the domain of universities and scientific societies and these organizations chose the areas they wished to explore in a haphazard way. Technology was primarily in the domain of industry.

Government's direct interest in science and technology was primarily restricted to certain aspects of health and national defense. Yet it also influenced science and techology in a number of indirect ways. For example, by maintaining a patent system, government helped foster an atmosphere that was beneficial to inventiveness, and in its procurement of technological products it indirectly gave some direction to the development of technology.

World War II changed government's role from a passive to an active one. Both the Allied powers and the Axis nations realized the importance of mobilizing science and technology for the war effort. Government became the major supporter of research, and scientists and technologists became the servants of government. This had a number of very significant implications both for science and government. For one thing, scientists became accustomed to doing the government's bidding, very often uncritically. While the exigencies of the war made this seem logical at the time, scientists today see this as unhealthy. Good science requires constant questioning of the established order. Slavish following of government directives by scientists and technologists carries with it the frightening flavor of an Orwellian world.

A second major outcome of the new relationship between scientists and government was that scientists developed a dependency on government support of their activities. Consequently, in future years, scientific work would be determined largely by the ups and downs of government research funding.

A final implication was that government became accustomed to working with scientists and technologists. They would now be given central roles in determining government policies in certain areas (especially in health and defense).

With the defeat of Germany and Japan, the prewar status quo did not return. True, scientists conscripted for the war effort now left government laboratories and projects and returned to their posts at the universities. However, things were not as before. Science and technology would no longer function in a laissez faire world.

Globally, two major postwar developments increased the role of planning and science policy in determining the direction taken by science and technology. One development was the Cold War, and the other the emergence of LDCs as actors to be reckoned with in the international arena. Each of these developments will be discussed briefly.

The Cold War

In the post-war era, the Cold War contributed to the maintenance of strong links between government and the scientific communities of the principal protagonists. The Cold War was fought on many different battlefields and the combatants were primarily concerned with winning the hearts and minds of uncommitted countries. Both the democracies and the Communist countries were out to show the superiority of their respective systems. Both sides placed great stock in demonstrating that they had superior scientific and technological capabilities; and scientific and technological performance became symbolic of national strength and worth. Thus, the launching of Sputnik by the Soviets in 1957 was a devastating blow to U.S. pride. It signified a loss of American prestige, a failure of the American way of life. The missile-gap crisis of 1960, where American cold warriors alleged that the Soviet Union had developed superior missile capabilities to those of the United States, furthered the sense that the United States was slipping in the technological realm.

In the United States, as a consequence of Sputnik, enormous quantities of funds were made available for scientific research. Funding of research increased dramatically until 1969-1970, at which time (during the height of the costly Vietnam War) it dropped.

Thus, the Cold War encouraged formulation of explicit science and technology policies. It was maintained that science and technology must be directed so as to increase national strength and prestige. This outlook prevailed in the Western democracies as well as in the Eastern bloc countries.

Emergence of LDCs

Many of the countries liberated from colonial domination in the 1940s, 1950s, and 1960s found they were in no condition to govern themselves effectively. This was particularly true in Black Africa, where decolonization occurred suddenly and on a large scale and where the colonial powers did little to prepare the indigenous populations for self-government.

It soon became apparent to the leaders of the new countries that development would not occur without planning. The LDCs did not have the luxury of growing in a trial-and-error fashion, as did Western countries, such as the United States. They did not have two hundred years to modernize. Pressures of burgeoning populations, limited resources, political instability, and general impoverishment dictated that the new governments work to bring some order to their countries as soon as possible, and this required major efforts of planning.

At the outset, planning activities simply entailed organizing effective government and establishing the rudiments of industry. The planning process involved the setting of development targets—which were often unrealistic—and the drawing of plans to meet the targets. In most LDCs, little attention was paid specifically to the scientific and technological content of the plans. However, science and technology were dealt with implicitly in the planning of certain broad sectors of society and the economy. The educational, health, agricultural, and defense sectors of the economy each had strong scientific and technical elements and these had to be addressed somehow. Individuals had to be trained in scientific and engineering disciplines, a cadre of health workers had to be developed, soils had to be tested, and so on.

By 1970 this lack of specific consideration of the role of science and technology in development changed. Development plans began to deal explicitly with science and technology. There are a number of explanations for this changed outlook. First, and perhaps most significant, the LDCs began to view technology as an engine of growth. While technology may not be a sufficient condition for development, it certainly is a necessary one. LDCs needed advanced technology in order to build economically strong societies and to produce goods that would be competitive in world markets.

Second, the LDCs felt they were in a state of dependency on the advanced countries so long as they lacked their own scientific and technological capabilities. In their view, the master-servant relationship between Northern and Southern countries would not end until the Southern countries could stand on their own in all respects.

Third, many LDCs recognized that they were at a disadvantage in dealing with the multinational corporations owing to their lack of technical expertise. LDCs needed the capital and technology that the MNCs could provide them; however, acquisition of their goods and services occurred according to the terms MNCs dictated. By developing indigenous scientific and technological skills, the LDCs believed they could increase their bargaining position in dealing with the MNCs and the countries of the North.

A tangible indication of the need to include provisions for science and technology in national development plans is found in the *World Plan of Action for the Application of Science and Technology to Development,* published in 1971.[1] One of the provisions of the plan of action was that LDCs should invest 1 percent of their GNPs to strengthen their scientific and technological capabilities. Throughout the 1970s, there was a great deal of international activity to alert LDCs to the importance of science and technology for development, as well as to help them integrate science and technology into their development plans. This activity culminated in 1979 with the U.N. Conference on Science and Technology for Development in Vienna.

Elements of Science Policy

Science policy contains two elements that are universal: (1) providing for adequate scientific and technical manpower levels to meet national needs; and (2) providing for sufficient fiscal resources to maintain an adequate scientific and technological infrastructure. In short, any science policy must take into account people and funds.

Before we discuss these two basic elements of science policy in detail, we should point out that, as governments increasingly recognize the importance of strong scientific and technological capabilities for a country to compete in international markets, they are discovering creative ways of affecting scientific and technological growth outside of the conventional approaches discussed in the following paragraphs. For example, governments can carry out science policy objectives through their procurement policies. Japan has until recent times helped underwrite the R&D of its domestic telecommunications industry by excluding foreign firms from sales of telecommunications equipment to Japan. Governments also have a dramatic impact on domestic scientific and technological developments through their regulatory tax, and fiscal policies, as well as through direct subsidies of target industries.

Manpower

Manpower planning is primarily concerned with meeting manpower needs through education and training, although these needs can also be met in other ways, for example, by importing technical talent from abroad. Education and training occurs on many different levels.

Primary and Secondary Schools. It is important that any society wishing to develop and maintain strong technical capabilities have decent primary and secondary schools. This is as true in an advanced country, such as the United States, as in a less-developed country. In the United States today there is a great deal of soul-searching about declining academic performance in its universities, and a consensus has emerged that the decline is directly attributed to a weakening of primary and secondary school education.

Post-Secondary School Education. A technology-based society must have an ample pool of skilled technicians such as electricians, machine tool workers, laboratory technicians, and mechanics. These individuals are the lubricant that keeps a technology-based society functioning smoothly. One of the great bottlenecks to the development of many Third World countries is the lack of necessary technicians because of a disdain for manual labor. Technicians can receive much of their education in secondary schools. However, many of their specialized training needs may best be met after they graduate from secondary school. In the developed countries, particularly in the United States, a large network of postsecondary schools exists to address specialized training requirements and help a country develop the technicians it needs.

University Training. Basic training in science and technology is provided today in universities. Training in the pure sciences is offered in conventional universities, while technological training is acquired principally at engineering schools. In smaller or poorer countries, it is common to have students receive their university training abroad in countries with first-class facilities.

Post-Graduate University Training. Individuals who will engage in advanced scientific and technological work generally continue their education beyond the undergraduate university level. In graduate school they focus their attention on rather narrow concerns and develop highly specialized expertise. Maintenance of adequate graduate school facilities in the sciences and technology is very expensive, since it entails the purchase of expensive specialized equipment and is highly labor intensive. Consequently, the best graduate schools are situated in highly affluent countries that act as magnets, attracting talent from all over the globe.

Adult Education. The Soviet Union has long recognized that education does not end with one's youth, and adult education on a large scale has existed in the USSR for decades. In the West, where emphasis has been placed on traditional, formal kinds of education, recognition of the merits of adult education has been more recent. The need for adult education in a society is obvious in a rapidly changing world: the material we learn in our youth soon becomes obsolete. Furthermore, technology puts society into a constant state of flux, and a highly mobile society requires that its citizens be able to move efficiently from one line of work to another. The emerging importance of adult education in the United States is seen in the fact that it is the fastest growing educational area in the early 1980s.

Manpower Planning Process

Manpower planning typically involves three steps.

Step 1: Assess Needs. In the first step, planners assess manpower needs, both for today and the future. Precisely how the needs assessment is undertaken depends on the specific situation facing manpower planners. The assessment may be highly formalized, the consequence of a careful survey performed by experts both in and out of government, or it may be very informal, based on nothing more than the gut feeling of some political leaders.

In an assessment of future needs, the figures are conjectural. The more distant the future, the more speculative the estimates, because in a rapidly changing world it is difficult to say with any precision what life will be like in five, ten, or more years.

Step 2: Determine How to Meet Needs. After manpower needs have been determined in step 1, planners must determine how they will be met. Exactly how this is done is again contextually determined. For example, it may be determined in step 2 that a country will have a serious shortage of electrical engineers in five to ten years. There are a number of ways the planners may choose to address this potential shortage. One might be to expand existing training facilities so that more electrical engineers can be educated. Another might be to send students abroad for training, rather than expand domestic facilities. Yet another approach might be to encourage students to pursue an education in electrical engineering by issuing ample scholarships to be applied to domestic universities with underutilized training capacity. The point here is that there are many different ways of approaching the problem and the precise way the planners choose will be largely determined by the facts of the situation.

Step 3: Implement Procedures for Meeting Needs. Once planners determine what procedures they will follow to meet their country's manpower needs, they have to implement them. That is, they may direct relevant organizations to develop new training programs, they may release funds for the new programs, and so on. It is not enough simply to implement their programs. They must also monitor them carefully to determine whether they are being carried out effectively.

Resources

As stated earlier, one of the universal elements of science policy is manpower planning, and the other is resource planning and allocation. For the most part, the chief resource we are concerned with is funding.

In all countries, funds for R&D activities are scarce. Even a very ample national R&D budget, such as exists in the United States, is insufficient to cover all the projects that should be supported. Thus, science policy analysts must allocate scarce resources among various competing uses.

These resources can be allocated in many different ways. They can be directed toward government laboratories (intramural support), as in the case of the laboratories of the Council on Scientific and Industrial Research (India), the National Institutes of Health (United States), or the USSR Academy of Sciences (USSR). Or else they can be directed toward scientific and technological endeavors outside the government (called extramural support). This second option can take many forms. For example, government may provide subsidies to key industries in order to strengthen desired scientific and technological capabilities. One of the best known examples of this is Japanese government subsidies and direction in the domestic computer industry. Another example is the giving of grants and contracts by government to research performers.

Range of Science Policies

There are great variations in how countries carry out science policy. In some countries, policy is directed only at relatively limited concerns (for example, health, defense, energy), while in others it is all encompassing (including, for example, industrial and commercial concerns). In some countries, it is just a peripheral interest of government, while in others it lies near the heart of government's activities. In some countries, science policy is carried out in a principal ministry, in others it is highly decentralized.

Dramatic differences in science policy approaches can be seen by comparing the world's two scientific and technological giants: the United States and the USSR. Their science policy styles closely reflect their economic, political, and social philosophies. In the United States, scientific and technological decision making occurs in a laissez faire environment, while in the USSR, all aspects of science and technology are carefully planned. Other countries with advanced scientific and technological capabilities generally fall somewhere between these two policies.

Science Policy in the United States

There are two distinguishing features of U.S. science policy: the process is highly decentralized; and it entails occasionally intense adversarial relationships between the policymakers.

These features are clearest at the highest levels of policymaking and become somewhat blurred at lower levels. Individuals familiar with the American political, economic, and social system recognize that these features characterize practically all types of public policymaking efforts in the United States and are not unique to the realm of science policy.

The Decentralized U.S. Decision-Making Structure

It can be argued convincingly that there is no science policy in the United States. Certainly, no unified command system exists for making and executing science policy decisions. The United States does have a number of departments that issue guidelines and policies in their areas of concern, such as the Departments of Agriculture, Commerce, and Defense, but there is no central Department of Science and Technology and no uniform guidelines and policies for science and technology.

Although there may not be a single cohesive science policy in the United States, policy decisions that have an impact on the course of the national development of science are constantly made throughout the research system. Some of these decisions bear directly on science, such as decisions on the allocation of funds for research. The effects of others are somewhat less direct, such as decisions pertaining to the maintenance of the public health system.

The focus here will be on science policy that originates in the federal government, although significant policy can be generated by other sectors as well. State and local governments, for example, make policy decisions routinely on educational and environmental issues that can have a measurable impact on how scientific research progresses in the United States. Likewise, the academic- and private-laboratory communities can devise their own policies, as they did with the issuance of self-imposed guidelines for research on recombinant DNA in the late 1970s.

Within the federal government, each of the major branches—executive, legislative, and judicial—plays a role in science policymaking. The executive branch, headed by the President of the United States, decides what areas should receive material support; supports research in the academic and industrial sectors; engages in research in its own laboratories; and promulgates regulations that may foster or hinder the development of American science. The operating arms of the executive branch that make science policy include various agencies (such as the National Institutes of Health, the National Science Foundation, and the Department of Defense), the Office of Management and Budget, and the Office of Science and Technology Policy, headed by the president's science advisor.

The legislative branch—composed of the Senate and the House of Representatives—is concerned with approving a federal budget and making laws for the nation. In its budget-making capacity the legislative branch can follow executive branch guidelines on how funds should be allocated among scientific programs, or it can ignore them and allocate the funds according to its own desires. As the national lawmaker, the legislative branch can introduce laws that institutionalize support of research areas that are perceived as especially important to the nation (for example, the passage of the National Cancer Act in 1971). It can also pass laws that affect the research climate (for example, patent and tax laws). Organizationally, the Senate and House operate through committees and subcommittees that specialize in various functions (for example, the Senate Appropriations Committee, the House Subcommittee on Science, Research, and Technology). In addition, the legislative branch obtains advice on scientific and technological affairs from its Office of Technology Assessment, the General Accounting Office, and the Congressional Research Service.

Even the judicial branch—composed of the Supreme Court and the federal appellate courts—has made decisions that have significant policy implications for science. A recent example of this is the Supreme Court's ruling on 16 June 1980 that new life forms created by scientists can be patented. However, the role of the judical branch in science policy formulation is generally oblique and is given little consideration in studies of U.S. science policy.

Some of the principal science-policy actors in the executive and legislative branches are listed below. This list helps make clear the extent to which science policy formulation is decentralized in the United States.

Executive Branch

Agency for International Development
Department of Agriculture
Alcohol, Drug Abuse, and Mental Health Administration
Commerce Department
Defense Department (including Naval Research Lab, Air Force Office
    of Scientific Research, etc.)

Energy Department
Environmental Protection Agency
Geological Survey
National Aeronautics and Space Administration
National Bureau of Standards
National Institutes of Health
National Oceanographic and Atmospheric Administration
National Science Foundation
National Technical Information Service
National Telecommunication and Information Administration
Office Management and Budget
Office of Science and Technology Policy
State Department

Legislative Branch

Senate Committees
Agriculture, Nutrition, and Forestry
Appropriations
Armed Services
Commerce, Science, and Transportation
Energy and Natural Resources
Environment and Public Works
Finance
Foreign Relations
Government Affairs
Labor and Human Resources

House Committees
Agriculture
Appropriations
Armed Services
Foreign Affairs
Government Operations
Interior and Insular Affairs
Merchant Marine and Fisheries
Science and Technology

Other Legislative Organizations
Joint Economic Committee
Office of Technology Assessment
Congressional Research Service
General Accounting Office

Adversarial Nature of U.S. Science Policy Formulation

Science policy formulation in the United States entails a great deal of give and take among individuals and organizations holding opposing viewpoints. Generally, the policy debate occurs in full public view, and the final outcome can be affected by the demands of scientists, public-interest groups, industry, and raw public-opinion. We will briefly examine this adversarial characteristic of science policy in the United States at both high and low levels of decision making.

At the highest level of policy formulation, the president may initiate a course of action. An example of such a presidential initiative was President Carter's decision, announced in April 1977, to abandon the development of breeder reactors in the United States on the grounds that it is not now facing a uranium shortage, and that by the time it does early in the next century, today's breeder technology will be obsolete. Carter specifically determined that efforts to build a breeder reactor at Clinch River, Tennessee, should be halted.

The translation of this initiative into an operational national policy required concurrence on the part of the legislative branch. Congress could have actively supported the president's initiative and the executive and legislative branches could have together made a great show of supporting a given science policy, as they did with the war on cancer in the early 1970s. Or else, Congress could have passively accepted the president's initiative and simply allowed funding of the Clinch River breeder to lapse. Instead, Congress took a stand against the president. Its position was influenced in part by a report it requested from the General Accounting Office, which questioned the validity of the data upon which the president based his decision. In addition, the resistance to the presidential initiative was seen as reflecting a pronuclear sentiment in Congress, which stood in marked contrast to Carter's antinuclear position. At any rate, the legislative branch overruled Carter's initiative and voted to continue funding of the Clinch River breeder reactor program.

Science policy initiatives can be made at high levels by Congress as well as by the president. A recent example was a House-initiated bill to authorize expenditures of twenty-five million dollars to engage in exploratory R&D on a Solar Power Satellite (SPS), which would beam microwave energy to earth receiving-antennas for conversion to electrical energy. The fully operating SPS system would cost an estimated five hundred billion to one trillion dollars.

This bill also demonstrates the adversarial nature of science policy formulation in the United States. Though it must be approved by the Senate and the president before it can become law, for years the Senate failed to act on the bill (and its predecessor), because of dissatisfaction that was thought to be based on environmental and financial grounds. Without Senate and presidential action on the bill, a commitment to support accelerated development of the SPS system cannot become national policy.

At lower levels of authority, a modified adversary system exists, characterized by bureaucratic decisionmaking. It is doubtful that at this level the American decisionmaking system is substantially different from that of most other developed countries.

Bureaucratic decisionmaking on science policy issues can have both fragmenting and consensual aspects to it. The former is illustrated by what is termed bureaucratic infighting. For example, two offices within an agency—one supporting basic research, the other supporting applied—may engage in a bitter struggle to assure that the agency increases its support of one or the other type of research.

Consensual decision making is typified by the common practice of circulating a policy memorandum through an office or agency in order to obtain comments that will lead to revisions. The memorandum may go through several iterations of review before it is completed, the rationale for this process being that many heads are better than one. In practice, the reviewers of the memorandum expunge its more controversial points so that the resulting document may possess little true initiative or original thinking. A typical example of this was the U.S. national paper presented at the U.N. Conference on Science and Technology for Development (UNCSTD), held in Vienna in the late summer of 1979. This paper purportedly outlined U.S. policy on science and technology for the development of LDCs. Drafts of it were circulated several times thorugh various government agencies, and in the process virtually all controversial points—some real, some imagined—were removed. The resulting document submitted to UNCSTD was so totally devoid of substance that it was useless as a policy statement.

Science Policy in the USSR

While science policy formulation in the United States is highly diffuse and informal, in the Soviet Union it is not. The USSR is a centrally planned country, which is very clearly reflected in its science policy. Soviet science policy formulation is strongly formalized. That is, policy is made by following clearly defined procedures. Furthermore, the major thrusts of policy formulation come from the top of the national leadership pyramid, where broad scientific and technological goals are established and passed down through the system for implementation. This helps to create a certain consistency in Soviet science policy that is lacking in the United States.

Soviet Research Performers

In the USSR, R&D activities occur under the auspices of the USSR Academy of Sciences, industrial ministries, or institutions of higher education. Because of the nature of Soviet communism, all R&D efforts, regardless of where they are undertaken, constitute government-supported research.

The USSR Academy of Sciences is the premier performer of basic research in the Soviet Union. Research is carried out in academy laboratories that attract the best graduates of scientific and engineering programs. Republic academies (for example, the Moldavian SSR Academy of Science) carry out research of interest to the local region and they do this under the guidance of the USSR Academy of Sciences. The academy laboratories not only carry out basic research, but undertake some development and applied research activities as well.

Research performed by the industrial ministries (for example, chemical industry; electrical equipment industry; tractor and agriculture-machine building industry) serve the needs of specific industries and tend to focus on development efforts and applied research, although relevant basic research may also be undertaken. Ministry-supported R&D is conducted in branch organizations and to a lesser extent in enterprise laboratories. The branch organizations tend to focus on specific industry-wide problems within the entire R&D spectrum, from basic research to the building of prototypes. Their work is utilized by many production facilities in the industry. Enterprise laboratories are situated in specific enterprises. Generally, the enterprise laboratory facilities are not as sophisticated as those of branch organizations. This emphasis on central research-laboratories servicing the industry as a whole through branch organizations leads to a separation of R&D from production that does not exist in the United States.

Soviet institutions of higher education play a far smaller role in research activities than those in the West. In the USSR their primary purpose is to teach and their research role is clearly secondary. Research performed in Soviet institutions of higher education is both basic and applied, the latter often being performed under contract to an industrial ministry.

Major Soviet Science-Policy Actors

Systems that are centrally planned and controlled tend to be complex and highly bureaucratized. They must deal with all sorts of contingencies, which leads to a proliferation of rules and of organizations to compose, implement, and monitor these rules. Thus it is that, while Soviet science policy is centralized, its formulation requires inputs from a wide variety of bureaucratic organizations. The centralized aspect of Soviet science policy derives from the top-down nature of decision making rather than from the number of organizations contributing to policymaking.

The principal actors involved in Soviet science policy are:

State Committee for Science and Technology. The SCST, created in 1965, takes a broad overview of scientific and technological activities in the USSR, and both initiates policy and coordinates national R&D efforts. Over the years, the SCST has played an increasingly important role in formulating national science policy, which is seen as reflecting a concern among the Soviet leadership that R&D is insufficiently integrated into the economy. The SCST is also charged with helping to design a role for applied research in the national five-year plans.

GOSPLAN. GOSPLAN is the Russian acronym for the State Planning Committee, the principal planning agent in the USSR. GOSPLAN attempts to integrate scientific and technological efforts into the overall economic system. Thus, its mission in regard to science and technology is similar to the organizational mission of SCST. Like SCST, it contributes to formulating policy for applied research in the five-year plans.

USSR Academy of Sciences. In the USSR, the academy is a very prestigious body. It was mentioned earlier that academy laboratories are the principal performers of basic research in the Soviet Union. In addition to performing research, the academy serves as the major policymaking body dealing with basic research in the country. It is charged with contributing to policy formulation for basic research in the five-year plans.

Council of Ministers of the Supreme Soviet. The Supreme Soviet is the highest-level legislative body in the Soviet Union. It is a large and cumbersome organization and chiefly rubberstamps policy dictated by the Communist Party Central Committee. Its executive arm is the Council of Ministers, composed of some ninety individuals, many of whom run ministries that rely on scientific and technological undertakings. The Council of Ministers in turn is governed by a Presidium, composed of a chairman, a first deputy chairman, and eleven vice-chairmen. Among the vice-chairmen are the heads of the SCST and GOSPLAN.

Central Committee of the Communist Party. The Central Committee of the Communist Party (CPSU), the highest economic and political authority in the Soviet Union, is composed of roughly 290 members and 140 candidate members. Its executive body is the Politburo, headed by the first secretary. Most of the Politburo's dozen or so members have technological backgrounds, having been trained in engineering and scientific areas. The party's directives dealing with S&T are transmitted to the vast Soviet R&D system in a number of ways. For one thing, the CPSU has its own Department of Science and Education, which advises it on scientific and technological affairs, makes policy, and issues directives to S&T agencies. Second, the CPSU transmits its directives through party cells that are found in research institutes. These party cells can have a major effect on the actions of institute directors. Finally, the CPSU is able to fill key positions in the R&D system with approved personnel. Thus, many individuals in leadership positions in the R&D community owe their jobs (and continued security) to the party.

Committee for State Security. The Committee for State Security (KGB) is the secret police organization of the Soviet Union. It maintains and monitors ideological purity in the nation. Its impact on scientific and technological activities is felt largely through its cadre of some seventy thousand full-time censors who oversee published material.

Advisory Councils. There are some forty councils in the SCST and some two hundred affiliated with the USSR Academy of Sciences. The councils are comprised of industrial leaders, research specialists, and scientists. They are the principal mechanism for obtaining grass-roots input into the decision-making process. Council members are highly regarded for their expertise, and their contributions to science policy formulation can be substantial.

The Planning Process

The basic planning instrument in the Soviet Union is the five-year plan. As a first step in devising this plan, the Communist party defines broad targets to address. GOSPLAN then begins incorporating these guidelines into a national plan. In the S&T area, some two hundred basic scientific and technological problems bearing on the economy are identified and programs are developed to meet them. The SCST selects head agencies to play a lead role in developing the specific details of the programs. Each head agency makes the program assigned to it operational: it assigns specific tasks to R&D performers, establishes deadlines, and generally defines tasks. The refined programs are then considered by GOSPLAN, the SCST, and the USSR Academy and included in the S&T section of the draft five-year plan. Their work is then forwarded to the Council of Ministers, the Politburo, and the Supreme Soviet for final approval.

Military R&D in the USSR

Military R&D efforts are very substantial in the Soviet Union. It is frequently noted in the West that while the quality of civilian research in the USSR is uneven, military technology is generally first-rate. One reason for this is that defense matters receive top priority in the USSR, often at the expense of the civilian sector.

Most military R&D efforts are carried out either by research institutes in the nine defense-related industrial ministries or in the USSR Academy of Sciences laboratories. The organization that attempts to coordinate military R&D and industrial activity is the Military-Industrial Commission.

Note

1. U.N. Advisory Committee on the Application of Science and Technology for Development, *World Plan of Action for the Application of Science and Technology to Development* (New York: United Nations, 1971).

Science, Technology, and International Organizations

9

Introduction

Modern international organizations have their origins in nineteenth-century Europe. Interestingly, from the perspective of this book, they came into existence in order to help countries deal with the resolution of international technical issues.

The first modern international organization was the Central Rhine Commission, which was established in 1804 by Germany's present-day predecessors and France. The commission was designed to deal with regulation of river traffic, maintenance of navigation facilities, and adjudication of conflicts involving violations of commission rules. A similar organization was created by several countries in 1856 to govern activities on the Danube River. This second organization was called the European Danube Commission. Both the Rhine and Danube Commissions still function today.

Also significant in the evolution of modern international organizations was the establishment of the International Telegraphic Union in 1865 and the International Postal Union in 1874. Both these organizations tackled problems of communication across national boundaries. The clear-cut successes of these two organizations stimulated other international cooperative efforts in health, agriculture, railroads, patents, and so forth.

Today international organizations abound. A major motivation for their creation is the need for some kind of forum to deal with problems that extend beyond the borders of a single country. Many of these problems have scientific or technological roots. Thus, we now have organizations that deal with pollution of the seas and air, use of outer space, communication across national boundaries, et cetera.

For the most part, the activities of international organizations are of peripheral interest to individuals engaged in international business. There are notable exceptions to this, however. For example, the activities of the International Telecommunications Union (ITU) regarding allocations of the radio spectrum are taken seriously. If governments and individual organizations did not abide by ITU guidelines, the international telecommunications system would be a shambles.

In this chapter we will very briefly identify some of the more salient international actors in the area of S&T, in order give the reader a fuller appreciation of the organizational and legal dimensions of international S&T and their relation to international business.

The UN System

Many of the best known international actors dealing with scientific and technical issues are affiliated with the United Nations. Some of these have missions that are directly oriented toward science and/or technology (for example, World Health Organization). In other cases, concern for S&T is peripheral to the central mission of the organization (for example, the U.N. Conference on Trade and Development). We will here briefly touch upon the more visible UN-affiliated organizations.

United Nations Educational, Scientific, and Cultural Organization

As its name implies, the United Nations Educational, Scientific, and Cultural Organization (UNESCO) has an interest in science. UNESCO was established in 1945-1946, has nearly 150 national members, and is head-quartered in Paris. Its scientific efforts are undertaken primarily through its division of natural sciences. In recent years, most of UNESCO's efforts have been directed at the question of the application of S&T in the Third World. In this context, it has paid special attention to science policy, scientific and technological education, natural resources, and the environmental sciences.

In furtherance of its objectives, UNESCO supports various colloquiums that address topics of interest and makes heavy use of outside experts and non-governmental organizations. It supports such a large number and wide variety of activities that by fiscal necessity each receives only minimal support. It is generally acknowledged that the shotgun approach used by UNESCO to carry out its mission has resulted in few meaningful results. Thus, while UNESCO is one of the few international organizations in the world explicitly concerned with a broad range of S&T issues, its impact is minimal.

International Telecommunications Union (ITU)

The ITU is one of the oldest international organizations in existence. It had its origins with the establishment of the International Telegraphic Union in 1865. In its first meeting, the International Telegraphic Union adopted the Morse code as the standard for international telegraphic communication.

The ITU has headquarters in Geneva and approximately 140 national members. Today, it is concerned with pursuing a number of objectives, chief among which are: allocating radio frequencies and registering frequency assignments; coordinating attempts to avoid radio frequency interference in broadcasting in different states; pursuing efforts to reduce charges for international telecommunications services; undertaking studies and collecting data for the benefit of its members; and promoting the establishment of telecommunications capabilities in LDCs. More will be said about the ITU in chapter 10, where we discuss the recent workings of the ITU's World Administrative Radio Conference.

UN Environment Programme (UNEP)

UNEP was created as a direct outgrowth of the U.N. Conference on the Human Environment held in Stockholm in 1972. Its headquarters are in Nairobi, Kenya.

UNEP's principal functions are to coordinate the environmental activities of governments and other international organizations; to alert governments and environmental organizations to issues that are important but which they have not yet considered; and to initiate new activities that are not likely to be undertaken by other organizations. A large part of UNEP's energy has been dedicated to monitoring the environment under the auspices of its Earthwatch Program.

One of the many difficulties UNEP has faced since its inception is determining precisely on which environmental problems it will focus. For example, the industrial nations would like to pay special attention to such things as pollution of the oceans and to conservation, while the Third World wishes to interpret the environment much more broadly, to include human welfare planning in the most general terms.

World Intellectual Property Organization (WIPO)

WIPO was brought into the UN system in 1974, has over eighty national members, and is located in Geneva. Prior to the 1970s, it operated independently and was known by the French acronym BIRPI (in English: Bureaux for the Protection of Intellectual Property). WIPO's chief function is to serve as the administrative guardian of the Paris Convention for the Protection of Industrial Property. That is, WIPO attempts to coordinate global patenting activities. It also serves as a consultant to LDCs that wish to establish domestic patenting systems. In recent years, WIPO has been viewed with displeasure by many critics from the Third World who see it as primarily interested in serving the status quo. In the view of these critics, the current international patenting system is a significant obstacle to the efficacious transfer of technology to the Third World. More will be said about LDC complaints regarding the Paris Convention in chapter 10.

World Health Organization (WHO)

WHO began operations in 1948, has more than 150 national members, and is headquartered in Geneva. It is concerned with major global health problems and in recent years, has focused heavily on health in the Third World. For example, in the 1970s it launched a major effort to combat shistosomiasis, a parastic disease afflicting and debilitating millions of people in certain Third World countries, as well as five other prominent tropical diseases. Perhaps WHO's best known accomplishment to date was its central role in finally ridding the earth of small pox, a scourge of mankind for millenia. As part of its anti-small pox strategy, WHO saw to it that any reported outbreak of the disease was immediately contained in a small area. Inasmuch as diseases afflicting people in tropical areas are not of much conern to the great medical research institutes of the industrial countries, WHO serves an important function in focusing attention on controlling and curing illnesses that otherwise would be ignored by the world scientific community.

World Meteorological Organization (WMO)

WMO is headquartered in Geneva and has approximately 150 national members. Its origins go back to the late nineteenth century, when it was established as a non-governmental organization called the International Meteorological Organization. It became a fully intergovernmental organization in 1946, at which time its name was changed to the World Meteorological Organization.

Initially, WMO was concerned only with establishing uniform standards for meteorological observations throughout the world. As time passed and great technological advances occurred, WMO's mandate broadened to include undertakings aimed at improving technological cooperation, training, education, and research; and looking at the interaction of man and his environment. Additionally, WMO established World Weather Watch (WWW) and the Global Atmospheric Research Program (GARP). WWW is a global meteorological network which links meteorological observations and information worldwide. GARP is a large-scale global research undertaking aimed at better understanding global weather fluctuations with a view to improving long-range weather forecasting.

Food and Agricultural Organization (FAO)

FAO was established in 1945, has over 120 national members, and is head-quartered in Rome. It was the first UN specialized agency to be created. Its basic purpose is to increase nutritional levels throughout the world by improving efficiency in the production and distribution of food and agricultural products. These activities clearly have a strong technical element to them.

Working alongside FAO is the International Fund for Agricultural Development (IFAD), which began operations in 1977 and is also head-quartered in Rome. It has more than 120 members. Its activities are directed at the world's poorest countries. IFAD's mission is to channel investments to LDCs to help them strengthen their commitments to food production, distribution, and storage, as well as to do research on nutrition and agriculture.

International Atomic Energy Agency (IAEA)

The IAEA was initiated as a consequence of a request by President Eisenhower and began operations in 1957. Its headquarters are in Vienna and it has nearly 110 members. Its overall objective is to promote the use of atomic energy for peace, health, and prosperity throughout the world. Subsidiary objectives include: to expedite the transfer of nuclear technology; to stimulate scientific and technological transfer of information; to assure that its advice is not used for military purposes; to help facilitate the training of scientists and technologists in nuclear technologies; and to establish international safety standards for the use of nuclear technology. The IAEA's promotion of nuclear technology extends beyond the obvious development of nuclear power, and includes, among other things, the use of nuclear technology for radio isotopes in medicine, agriculture, and industry.

UN Committees

Scientific and technological efforts in the United Nations frequently are pursued more vigorously in special committees than in large bureaucracies. Two recent examples of this are the Committee on the Peaceful Uses of Outer Space (COPUOS) and the Intergovernmental Committee on Science and Technology for Development.

COPUOS was responsible for drafting the Moon Treaty, which was approved by the General Assembly (with no vote) in December 1979, and which is discussed in more detail in chapter 10. It also served as the organizer of UNISPACE, the Second UN Conference on the Exploration of Peaceful Uses of Outerspace, which was held in Vienna in August 1982. In recent times, COPUOS's attention—via its legal and technical subcommittees—has been directed to examining remote sensing, direct broadcasting from satellites, the uses of nuclear power in outer space, the application of space technology to meeting the needs of LDCs, and the definition of what constitutes outer space.

The Intergovernmental Committee on Science and Technology for Development was created by the UN Conference on Science and Technology for Development held in Vienna in 1979. As its name suggests, the Intergovernmental Committee's mission is to coordinate and promote UN activities bearing on science, technology, and development. One of its principal tasks has been to raise large amounts of funds to support the application of S&T for development. While the participants of UNCSTD created a target of some $250 million a year to support such activities, the Intergovernmental Committee has been having difficulties in raising even a small fraction of this figure.

Non-technical UN Organs and Affiliates

Interestingly, the UN organ which has had perhaps the broadest and most visible impact on international scientific and technological undertakings is the UN Conference on Trade and Development, better known by its acronym, UNCTAD. Unlike the UN affiliates discussed above, UNCTAD is not a technical organization. Its mission is broadly economic: to foster the development of LDCs through trade policy. However, UNCTAD's LDC members recognize that technology is an engine of development, and consequently they afford it a high position in their listing of global priorities. Thus, we find UNCTAD making well-publicized demands for restructuring the international patent system; for establishing a code of conduct governing the transfer of technology from MNCs to LDCs; for declaring the resources of the oceans and outerspace the common heritage of mankind; and so on. UNCTAD's impact on these technology-based issues has been more psychological than tangible. That is, few of the demands have actually been resolved to UNCTAD's liking; however, by making the demands in the first place, UNCTAD has forced a dialogue between the countries of the industrial North and underdeveloped South on matters that never before received public attention.

Other significant UN-related organizations are the U.N. Development Program (established 1966, headquarters in New York) and the World Bank group (headquarters in Washington, D.C.). Both these organizations focus on the economic development of Third World countries. However, as in the case of UNCTAD, they recognize that technology is an engine of growth, and consequently there is a heavy technical component to their activities. The U.N. Development Program (UNDP) devotes a great deal of effort to coordinating and administering the technical assistance to LDCs provided through the U.N. system, while the World Bank group provides financial assistance for projects in LDCs that generally have technical roots.

Other UN Actors

Other UN actors that have an interest in scientific and technological affairs, but that have not been discussed here include the International Civil Aviation Organization, the Inter-Governmental Maritime Consultative Organization, the Universal Postal Union, and the U.N. Industrial Development Organization.

Organization for Economic Cooperation and Development (OECD)

The OECD has a membership of twenty-three countries that, taken together, constitute what is called the industrialized West (including Japan). The OECD, with headquarters in Paris serves as a forum for discussing the problems and challenges facing Western countries. It suggests approaches its member countries can take to address these problems and challenges. Initially, the OECD focussed exclusively on economic and trade issues. In more recent times it has expanded its outlook to include such things as social problems, informatics, the environment, natural resources, underdevelopment, and science policy.

The OECD deals with science policy issues through its Committee for Scientific and Technological Policy and its Directorate for Scientific Affairs. It supports studies, collects data, and disseminates information in many different areas. For example, as a substantial portion of its activities are directed at examining the scientific and technological capabilities of its member countries, it collects uniform statistics on these capabilities, with particular attention to industrial R&D efforts.

The OECD also looks at the science policies of member countries and investigates environmental issues of interest to them; in recent years, it has delved into the role of information systems in industrial societies and has investigated the role of S&T in the Third World.

The studies and conferences supported by OECD are designed to help member countries in their decision making on scientific and technological matters. The consensus seems to be that OECD's S&T efforts are perhaps of some use to the smaller OECD countries (such as Sweden), and to those that are not fully industrialized (such as Spain); but that they are largely ignored by the larger countries. The reason for this is that the smaller countries are able to use OECD projects to supplement their own limited scientific and technological policy activities, whereas the larger countries have sophisticated science policy staffs who better serve their needs than the OECD. Perhaps the greatest benefit of the OECD's S&T projects to the larger countries is that they enable their science policy decision-makers to maintain open communication channels to each other.

International Institute of Applied Systems Analysis (IIASA)

IIASA is a product of the Cold War and represents an attempt to bridge some of the hostilities between East and West through international scientific collaboration. It began operations in a Vienna suburb in 1972, at the height of U.S. and Soviet attempts to reduce mutual antipathy through detente. Its membership is composed of academies of science and equivalent organizations of slightly fewer than twenty countries from East and West.

IIASA operates by having scientists and engineers visit its Vienna headquarters for periods typically ranging from several weeks to one or two years. These international visitors then work on problems that have a bearing on the functioning of modern societies. The nature of these problems has varied with time, reflecting perceived changes in problems facing modern societies. Typically, problems are identified with areas such as energy, urbanization, food and agriculture, technology management, and decision analysis.

The principal mechanism for disseminating the findings of the IIASA projects to the public is through the publication and widespread distribution of reports.

Because IIASA was very much a child of detente, it was not surprising to find that as the fortunes of detente began to wane, IIASA's political *raison d'etre* would come under scrutiny. Thus, when Ronald Reagan became president, IIASA became an instrument to demonstrate U.S. dissatisfaction with Soviet policies. This took tangible form when the U.S. withdrew its contributions to IIASA (which had been substantial) in December 1982.

International Council of Scientific Unions (ICSU)

ICSU is an international nongovernmental organization composed of eighteen scientific unions and about seventy-five national members and associates. Each scientific union focuses on some disciplinary area. For example there is an International Astronomical Union, an International Union of Goedesy and Geophysics, an International Union of Biological Sciences, an International Mathematical Union, and so on.

ICSU's expressed objective is to encourage international scientific activity for the benefit of mankind. Among its many activities are: the initiation, design, and coordination of international scientific research projects (the 1957-1958 International Geophysical Year is the best known of these); the organization of conferences, symposiums, meetings, and summer schools throughout the world; the production of publications, including journals, newsletters, handbooks, proceedings of meetings, symposiums et cetera; and the sponsorship of committees that focus on particular issues (e.g., the Committee on Space Research [COSPAR], which came into being in 1958; the Committee on Science and Technology in Developing Countries [COSTED], 1966; and the Scientific Committee on Problems of the Environment [SCOPE], 1969).

Operational Organizations

International organizations have been set up that, like IIASA, allow scientists and engineers from different countries to congregate and to tackle scientific and technological problems together.

Certainly one of the best known of such organizations is the European Center for Nuclear Research (CERN), the creation of several Western European countries. These countries have pooled their resources to support an expensive research facility to examine questions in high-energy physics. Research carried out at CERN is considered of first-class quality.

Another well-known organization is the International Centre for Theoretical Physics, located in Trieste and supported by the Italian government, the IAEA, and UNESCO. This center brings together top talent in theoretical physics from both the advanced countries and the Third World. It gives high-quality scientists from the Third World an opportunity to interact with fine Western minds in theoretical physics, and consequently enables them to overcome some of their problems of isolation from mainstream research.

Other well-known functioning international research operations include the International Rice Research Institute, located in the Philippines and focusing on research advances in the production of rice, a staple crop for much of the world's population; the International Center for Insect Physiology and Ecology, located in Nairobi, and bringing together scientists from LDCs and industrial countries to study entomology-related problems; and the International Maize and Wheat Improvement Center (CIMMYT), a Mexican-based research operation whose work resulted in the so-called Green Revolution that created very high-yielding varieties of wheat.

Notes

A number of books have been written that look at different aspects of science and technology and their relationships to international organizations. See, for example, A.J. Dolman, *Resources, Regimes, World Order* (New York: Pergamon, 1981); S. Brown, N. Cornell, L.L. Fabian, and E.B. Weiss, *Regimes for the Ocean, Outer Space, and Weather* (Washington, D.C.: The Brookings Institution, 1977); E.B. Haas, M.P. Williams, and D. Babai, *Scientists and World Order* (Berkely: University of California Press, 1977); M.J. Moravcsik, *Science Development* (Bloomington, Indiana: PASITAM, 1975); E.B. Skolnikoff, *The International Imperatives of Technology,* Research Series, no. 16 (Berkeley: Institute of International Studies, 1972); and R.A. Solo, *Organizing Science for Technology Transfer in Economic Development* (Lansing, Mich.: Michigan State University Press, 1975).

10 Third World Demands for Developed-Country Technology

Introduction

Background

In recent years the international business scene has been beset with a wide array of uncertainties: wildly fluctuating exchange rates, the emergence of a new mercantilism, political instability in less-developed and some Western countries, high levels of inflation in developed countries, the accumulation of large debts to pay oil bills, and so on. Enterprises engaged in international commerce have had to take stock of these uncertainties and accommodate them in their global strategies. For the most part, these uncertainties have been familiar developments on the international scene, having risen and subsided at various times in history. However, in recent times a new development has entered into the picture, one that may have a substantial impact on how international business is conducted, but which is poorly understood. This is the increase in demands by the LDCs for access to technology produced in the developed countries.

These demands are singularly lacking in modesty. They entail a restructuring of the international patent system; preferential access to Western technology; recognition that the international seabed, the moon and asteroids, and the airwaves cannot be exploited simply for the benefit of the developed countries; and the imposition of constraints on the way MNCs do business. They can be described as multibillion-dollar demands, since if they are met they will have the effect of shifting billions of dollars of know-how from the developed countries to the Third World, or else of placing billions of dollars of business activities out of reach of the developed countries' enterprises. To the Third World, these are just demands, one way in which the past injustices of the international commercial system can be righted. To Western ears, these demands sound strident. Particularly irksome to Westerners is the fact that the very individuals who make these demands at international forums are often members of the privileged elite of their countries, so that their claims against injustice have a hollow ring to them.

Unfortunately, the apparent extremity of the LDC demands and the corresponding muddled, defensive, and often hostile responses of Westerners combine to create a situation where the legitimate views of both sides have been obscured. Very little systematic thought has been given to this matter. It can safely be said that although many businesses come up against the LDC demands in one form or another, very few of them have developed any kind of an appreciation for the broad context in which they are made. In this chapter, some of the more salient LDC demands and their implications will be examined.

In order to appreciate the nature of current LDC demands for improved access to developed country S&T, it is necessary to examine two parallel, though not unrelated, developments. One is the increasingly active role played by the Third World in its attempts to improve the economic lot of the LDCs in the world economy. The other is the heightened awareness by LDCs that the acquisition of technology is a necessary, although not sufficient, condition for economic growth. We shall look at each of these developments in turn.

Third World Economic Militancy

Prior to the 1960s, the principal concern of the Third World was decolonization. While Latin American countries had gained independence from their colonial masters in the nineteenth century, and the Mideastern and Asian countries had acquired independence in the years following World War II, by the end of the 1950s Africa was still predominantly under colonial rule. The decolonization of Africa was a cause around which the whole Third World rallied.

The years 1960-1965 were crucial for decolonization in Africa. During this time, twenty-eight African countries gained independence. Thus, by the mid-1960s, the colonial experience was largely a thing of the past, with the notable exceptions of Angola, Mozambique, and Rhodesia. With the demise of colonialism, the former colonies, both in Africa and elsewhere, could turn their attention to what was called nation building and to providing their populations with adequate living standards. This proved to be a difficult undertaking, particularly in view of the cultural incompatabilities within the newly liberated countries, administrative deficiencies, capital shortages, technological shortcomings, and so on.

A major consequence of the widespread decolonization of the early 1960s was the swelling of the ranks of Third World countries in the United Nations, it specialized agencies, and other multilateral forms. Scores of countries, which previously had little or no voice in international assemblies, quickly came to dominate these assemblies in number. The liberated colonies were a diverse lot, representing wide-ranging cultures, languages, faiths, and races. They did, however, share a number of very important characteristics. First, they were all poor. Second, they were technologically backward, lacking both industry and the technological capability to maintain it. Third, their societies were generally traditional and their citizens lacked most of the educational skills necessary to sustain a modern nation. Fourth, they often had burgeoning populations that could barely be fed by domestically produced foodstuffs. The LDCs had a final characteristic in common that strongly colored their world outlook: most of them shared a common, humiliating colonial heritage and perceived themselves as exploited by the Northern countries.

With the establishment of the UN Conference on Trade and Development in 1964, a pattern of North-South confrontation at international forums was established that still exists today. UNCTAD was set up as an LDC counterpart to the General Agreement on Tariffs and Trade (GATT). GATT was established in 1947 to regulate international trade. Its objective was "the substantial reduction of tariffs and other barriers to trade and . . . the elimination of discriminatory treatment in international commerce."[1] However, in the early 1960s, the LDCs saw GATT as principally serving the interests of the developed countries, and so they pushed for the establishment of UNCTAD. UNCTAD has held a total of five plenary sessions: the first in Geneva (1964), the second in New Delhi (1968), the third in Santiago (1972), the fourth in Nairobi (1976), and the fifth in Manilla (1979). One of the major accomplishments of the first meeting was the establishment of a group of seventy-seven developing countries that would act in concert to represent the views of the Third World on international trade matters.[2] This collection of LDCs came to be known as the Group of 77 (or in a short-hand version, G-77). Today, its membership consists of considerably more than one hundred countries.

Several years were to pass before UNCTAD and the Group of 77 were to find their voices and to become forces that would be taken seriously in international affairs. What gave crucial impetus to G-77 activities and encouraged LDCs to assume a militant posture was the oil embargo of 1973-1974. The oil embargo, undertaken by the Arab members of the Oil and Petroleum Exporting Countries (OPEC), showed that when LDCs acted in concert they could have a substantial impact upon the actions and attitudes of the powerful and rich countries of the North. Thus, we find the Group of 77 calling in 1974 for nothing less than the establishment of a New International Economic Order (NIEO).[3] The basic argument behind the call for a NIEO was that the international economic system was clearly structured to benefit the wealthy countries, and that it was now time to restructure it to enable the LDCs to benefit. In restructuring the economic order, it would be necessary to give LDCs special preferences in order to compensate them for past injustices in the international system. A whole body of positions has come to be associated with the NIEO outlook, positions on matters such as trade in commodities, international shipping, international finance, technology transfer, and the behavior of MNCs.

Third World Awareness of the Need for Science and Technology

The first notable indication of an awareness that S&T played a significant role in national development was the convening of a UN conference on the Application of Science and Technology for the Benefit of the Less Developed Areas (UNCSAT) in 1963. One outcome of the conference was the establishment of the UN Advisory Committee on the Application of Science and Technology for Development (ACAST).[4] ACAST was composed of distinguished experts and was given the task of overseeing the UN's work that had a bearing on S&T.

Since the convening of UNCSAT a fair amount of international S&T activity has occurred relevant to LDCs. Some of the more visible activities pertaining to North-South interaction in the international arena are listed below.

1. UN Conference on the Application of Science and Technology for the Benefit of the Less Developed Areas (UNCAST), 1963
2. Establishment of the UN Office of Science and Technology (OST), 1963
3. Establishment of the UN Advisory Committee for the Application of S&T to Development (ACAST), 1963
4. UN Conference on Trade and Development (UNCTAD), 1964, 1968, 1972, 1976, 1979, 1983
5. Establishment of UN Development Programme (UNDP), 1965
6. Establishment of UN Industrial Development Organization (UNIDO), 1967
7. Publication of the World Plan of Action, 1971
8. Establishment of the Committee on Science and Technology for Development (CSTD), 1971
9. UN Conference on the Human Environment, 1972
10. World Population Conference, 1974
11. World Food Conference, 1974
12. Law of the Seas Conference (UNCLOS III), 1974-1982
13. UN Conference on Human Settlements (HABITAT), 1976
14. UN Water Conference, 1977
15. UN Conference on Desertification, 1977
16. UN Conference on Technical Cooperation Among Developing Countries (TCDC), 1978
17. UN Conference on S&T for Development (UNCSTD), 1979
18. Establishment of the UN Intergovernmental Committee on Science and Technology for Development, 1979
19. World Administrative Radio Conference, 1979
20. UN Conference on the Peaceful Uses of Outerspace (UNISPACE), 1982

Significant benchmarks in this listing are the publication of the *World Plan of Action* in 1971, the creation of the Committee on Science and Technology for

Development (CSTD) in 1971, and the holding of the UN Conference on Science and Technology for Development (UNCSTD) in 1979.[5] The *World Plan of Action* was a document put together—largely by Westerners—under the auspices of ACAST. It suggested S&T goals LDCs should set for themselves, as well as strategies for attaining them. CSTD was created at the behest of the LDCs because they were dissatisfied with ACAST's rather technical approach to the question of S&T for development. CSTD would have more of a political orientation and would practice UNCTAD-style confrontations with the developed countries. Finally, UNCSTD was held in Vienna in the late summer of 1979, after several years of detailed preparation on the part of all the participants. It is generally agreed that UNCSTD's chief value lay in increasing both developed country and LDC awareness of the problems and prospects of establishing scientific and technological capabilities in LDCs.

A new characteristic of international scientific and technological activities since the rise of G-77 militancy is the obvious intrusion of politics into what are purportedly technical proceedings. For example, it became de rigeur at international conferences for G-77 countries to include in the published proceedings condemnations of South African apartheid and the Israeli occupation of Arab lands, no matter how irrelevant the activities of these two countries were to the content of the conference. Other examples: At the UN Water Conference (March 1977), a resolution directed against the United States was passed by G-77 countries that expressed "earnest wishes" that a "just and equitable" solution would be forthcoming on the matter of the Panama Canal Zone, enabling the Republic of Panama" . . . fully to exercize its sovereign rights in the part of its territory known as the Canal Zone, and consequently to formulate a national policy for the full development of water resources."[6] The last phrase was a patently feeble attempt to relate the political condemnation to the subject matter of the conference. At the UN Desertification Conference (August-September 1977) an Israeli document on their experience with reclamation of the desert was condemned by G-77 countries as being an ideological polemic,[7] and in a vote which pitted G-77 countries and the Eastern block against the Western nations, it was determined that the document should be expunged from the proceedings.[8]

Major Third World Demands

The marriage of Third World economic militancy and Third World awareness of the importance of S&T for development led the LDCs to make rather bold demands for increased access to developed country technology. The demands were not backed with any tangible threats, since the LDCs were not arguing from a position of strength, but were principally based on moral considerations.

Inasmuch as the demands have been made in an institutional setting—primarily via UNCTAD and G-77—they are often associated with distinct institutional undertakings. Thus the UN Law of the Seas Conference, the World Administrative Radio Conference, the Moon Treaty, and the U.N. Conference on Science and Technology for Development have been forums in which many of the LDC demands have been articulated. Demands for a restructuring of the international patent system have also occurred in an institutional context, since the system is governed by the Paris Convention on the Protection of Industrial Property, which is administered by the UN's World Intellectual Property Organization. Even an issue such as the behavior of multinational corporations—which one would think would be handled individually by each country through, for example, its investment laws—has become institutionalized through attempts to create international codes of conduct governing the transfer of technology.

We will examine in detail three of the most significant demands by LDCs—for declaring the resources in the global commons as being the common heritage of mankind, for restructuring the international patent system, and for regulating the technology transfer activities of the MNCs—in their institutional contexts (respectively, the UN Law of the Seas Conference, the Paris Union, and UNCTAD's code of conduct for technology transfer). We will also more briefly examine three additional demands in order to give a fuller appreciation of the broad range of demands being made. Included in this second group are demands pertaining to allocations of the radio spectrum, the building of scientific and technological capabilities in LDCs, and the exploitation of the resources of the moon. The institutional settings in which these demands are being made are, respectively: the World Administrative Radio Conference, the UN Conference on Science and Technology for Development, and the Moon Treaty.

UN Law of the Seas Conference

The Third UN Conference on the Law of the Seas (UNCLOS III), which began deliberations in 1973 and concluded them nine years later, strove to develop a treaty pertaining to the law of the sea, its most important points being: (1) a redefinition of waters over which coastal states have jurisdiction and a clarification of navigation rights in these waters; and (2) the establishment of a regime for the exploitation of resources from the deep seabed in international waters. It was generally acknowledged that in respect to this second issue, whatever regime that emerged would have a substantial impact on the future formulation of international law.

Since the time of Hugo Grotius, in the early seventeenth century, national sovereignty extended three nautical miles out to sea from a country's coastline. The seas beyond the three-mile limit constituted international waters and were under no nation's domain. International waters were, in effect, a global commons. The first significant departure from traditional international sea law occurred in 1945 when President Truman declared that the United States had jurisdiction over the seabed resources of the continental shelf, which extended far beyond the three-mile limit.[9] In 1947 Chile, Peru, and Ecuador declared jurisdiction over both the seabed and waters that extended two hundred nautical miles out from the coastline. Over the next several decades, different countries unilaterally declared jurisdiction over waters extending various distances. It became clear that the matter of national jurisdiction over coastal waters was chaotic and that some attempt should be undertaken to apply a uniform standard for all coastal states.

Meanwhile, as technology led to increased mastery of the oceans, the feeling grew that some rules should be established for exploiting the oceans' resources. During the exploratory voyage of the H.M.S. Challenger in 1873-1876, manganese and iron oxide nodules were discovered lying on the ocean floor. At first these nodules were viewed principally as scientific curiosities. However, in the late 1950s John Mero, an American, showed that the nodules could have substantial economic value. After Mero published his findings, industry began entertaining the possibility of mining the seabed in earnest.[10]

Traditionally, in its attention to the high seas, international law has been primarily concerned with governing the behavior of ships navigating the surface of the oceans. Little attention focussed on the seabeds and their riches for a very obvious reason: the technology did not exist for exploiting them. However, by the mid-twentieth century this situation changed. The question naturally arose as to what were the rights of individuals, firms, and states in exploiting the resources of the global commons. This issue took on a special urgency when it was determined that a very rich concentration of nodules, in the Calrion-Clipperton area of the Pacific, lay under international waters.

The position taken by the United States was fairly enlightened, considering that it had far and away the greatest technological capability to exploit the deep sea. In 1966, President Lyndon Johnson declared: "We must ensure that the deep seas and ocean bottoms are and remain the legacy of all human beings."[11] The following year Arvid Pardo of Malta declared the resources of the deep seabed as being "the common heritage of mankind" in a now-famous speech before the UN General Assembly.[12] In 1968 the General Assembly established a Committee on the Peaceful Uses of the Sea-Bed and Ocean Floor beyond the Limits of National Jurisdiction. The work of this committee led to adoption by the General Assembly in December 1970 of a Declaration of Principles Governing the Sea-Bed and the Ocean Floor, and the Subsoil Thereof, beyond the Limits of National Jurisdiction.[13] With this declaration, the concept of common heritage of mankind gained a certain international legitimacy.

Although they would ultimately come to regret it, the developed countries of the West sought to link the issue of navigation rights on the sea with the matter of exploiting the resources on the deep seabed.[14] They reasoned that the Third World countries, which were enthusiastic about mining the oceans' riches, would be inclined to make concessions on navigation issues if they saw themselves prospering on the seabed mining issue. The U.S. defense establishment, in particular, was anxious to clarify the rules governing submarine passage through waters under national jurisdiction and navigation through straits. Thus, when UNCLOS III convened in New York in 1973 and in Caracas in 1974, it had on its agenda the resolution of certain navigation issues as well as issues concerning exploitation of the resources of the international seabed. (UNCLOS III would also turn its attention to two lesser issues: the marine environment and scientific research; and the establishment of a system for the compulsory settlement of disputes.)

Over the years a solid consensus developed on the clarification of navigation rules and in the establishment of uniform standards for defining national jurisdiction over waters. The territorial sea was extended to twelve miles from three. States would have full sovereignty over these waters, except that they must permit innocent passage of ships. Adjoining the territorial sea is the contiguous zone, which extends out an additional twelve miles. In this zone, the coastal state can impose its own customs, fiscal, immigration, and sanitary standards. Extending from twenty-four to two hundred miles out is the exclusive economic zone (EEZ). The coastal state has jurisdication over the waters, seabed, and subsoil in the EEZ. However, for maritime purposes the area should be viewed as international waters.

It is interesting to note that although the LDCs were instrumental in pushing for a two hundred mile EEZ, rich countries would be the principal beneficiaries of the new system. The ten countries with the largest EEZ jurisdictions are, in descending order: the United States, Australia, Indonesia, New Zealand, Canada, USSR, Japan, Brazil, Mexico, and Chile. Three of the four LDCs in this list are well on their way to industrializing. Among Third World countries not on this list, those with the largest EEZs are countries like Libya, Nigeria, Venezuela, and the Gulf states, which already have ample mineral wealth.[15]

Progress was slower on reaching an agreement on the mining of the deep seabed. The basic problem was that only a handful of companies from the United States, Europe, and Japan had the technology and financial resources to recover the manganese nodules from the ocean floor, and whatever agreement was achieved on mining rights would have to be palatable to these companies. They have to accept the risks and costs associated with the deep-sea endeavor and have to develop the needed technology.[16]

How much of their booty would they be willing to part with in order to meet some vague obligations regarding the common heritage of mankind?

Mining the deep seabed is an enormously costly undertaking. The nodules are typically located in a hostile physical environment some three miles below the surface of the sea, where the pressure is close to 7,000 pounds per square inch, there is no light, and it is very cold. One scale model of a fully operating nodule recovery system reaches the sea floor through a three mile pipe that is two feet in diameter and weighs approximately seven million pounds.[17] The nodules are actually gathered on the ocean floor by a remote-control robot that directs them into the pipeline, through which they are pumped to the surface. The technology and materials needed are far more advanced than for comparable operations on the earth's surface.

Not only would a deep-sea mining venture require huge capital outlays—it would also entail tremendous risks. One set of risks would be those attending any venture into a brand-new area where the unknowns of production are great. Will the mining technologies developed for the task be adequate? Will operating costs be low enough to make the venture profitable? Another set of risks would be tied to the price of the minerals being recovered. As with most commodities, mineral prices tend to fluctuate dramatically. Once the mining operations begin, will mineral prices be high enough to cover the costs of the undertaking? A final set of risks would be rooted in the lack of knowledge about the extensiveness of the mineral deposits. The current knowledge about the deposits is based on limited sampling techniques and there are great variations in estimates of the abundance of nodules on the ocean seabed. For example, by one estimate there are enough nodules in the Clarion-Clipperton area to support roughly twenty-eight mining operations, each producing some seventy-five million tons of nodules in its lifetime; whereas by other estimates there are only sufficient nodules to support up to eleven mining operations, each producing from three to four million wet metric tons for twenty years.[18]

Considering the costs, risks, and technology involved in the deep-sea mining of the nodules, it is easy to understand the reluctance of profit-making corporations to make any international agreements that would put a crimp on the returns on their investments. By the same token, the Third World position is also understandable. Why should these companies be entitled to profit from resources that do not really belong to them in the first place, but are part of the global commons?

Over the years, various proposals were entertained at UNCLOS III for an international regime on mining the deep seas. The Group of 77, for example, wanted the establishment of an all-powerful International Seabed Authority (ISA) which would have control over the mining operations of companies. An Enterprise would serve as the operating arm of the Authority in mining the seabed and would have the right of first refusal when it came time to selecting choice mining sites. Also, it was to be understood that the LDCs would receive special consideration in all aspects of the undertaking and that this should not be viewed as discriminatory.[19]

The Soviet Union proposed a system that emphasized the role of states in the regime. The mining activities in the area under exploitation would be conducted both by state parties and the Authority, with the Authority determining where it would conduct its operations. The Soviet proposal also put forward the view that special consideration should be given to LDCs and landlocked countries.[20]

Finally, the United States suggested the establishment of a parallel, or dual access, system. The Authority's Enterprise would stand on equal footing with other applicants for contracts to engage in mining operations. Contracts would be entered into if the applicants were qualified by virtue of their financial standing and technological capability. All contractors would be required to accept the supervision of the authority. As in the case of the G-77 and Soviet proposals, the United States suggested that special attention should be given to helping the LDCs.[21]

The bargaining that proceeded in the various sessions of UNCLOS III ultimately resulted in the acceptance of a parallel system, where applicants for mining operations would bring to the Authority descriptions of two sites of equal value where they would be willing to work. One site would be accepted by the Authority for exploitation by the Enterprise (either by itself or through a joint venture), while the second site would be assigned to the applicant. Since the Enterprise would be operating according to the concept that the deep-sea resources are the common heritage of mankind, the proceeds of its mining operations would be shared internationally, especially among the LDCs. In order to help capitalize the Enterprise, the developed countries would supply funds for an initial venture. LDCs would additionally benefit from the mining operations of the private contractors through a complicated system of fixed fees and royalty payments. Production would be limited at the outset of operations so as not to disrupt the economies of land-based producers of minerals.

An especially controversial provision in the draft treaty was one that requires contractors to transfer mining technology to the Enterprise if it is not available in the open market on reasonable terms. It is difficult to see how private companies, which developed their proprietary technology at great expense, would be willing to part with it so readily. On the other hand, it is equally difficult to envision the enterprise operating successfully without the needed technology.

Although a number of significant problems still needed to be ironed out, the general feeling at the conclusion of the ninth session of UNCLOS III in August 1980 was that the treaty would be signed by the Summer of

1981.[22] However, the election of Ronald Reagan in November 1980 created another major obstacle to the conclusion of UNCLOS III. Responding to strong criticism of the treaty in the United States,[23] the new administration announced at UNCLOS III's tenth session in the spring of 1981 that it was reviewing the entire draft convention and that the review would not be completed until early 1982. Since United States accession to the treaty was absolutely vital to its viability, UNCLOS III was effectively put into limbo for an indefinite period as a consequence of the unilateral American action. In May of 1982 a vote was finally taken on the treaty in the United Nations. The United States, along with three other countries, voted against the treaty. A total of 130 nations voted in favor of it, and 17 abstained. For the time being, then, navigation and deep-sea mining issues would remain unresolved.

Revision of the International Patent System

A substantial amount of attention has been drawn to the international patent system and its impact on the development of technology in the LDCs. The system as it now stands is viewed by G-77 as working against the Third World, and any fair appraisal of the situation would bear out this view. At the heart of the problem lies the balancing of the rights of inventors against the promotion of the public interest. In the developed countries, the rights of inventors and the public interest are to a large extent complementary. In LDCs, however, they are often at loggerheads.

Patents are granted for a number of reasons, two of which are particularly significant. First, it is hoped that through patents innovation will be encouraged and new products and processes will be introduced into the economy. Without protection good ideas could be pirated and inventors would not enjoy the fruits of their labor, which would have a negative impact on inventiveness and the willingness of inventors to take risks. Second, in granting a patent, governments require the inventor to disclose his invention, that is, to describe in a public document how it works. Public disclosure serves two basic functions: it reduces duplication of effort, since other inventors may determine through a patent search that their ideas have already been developed; and it stimulates further inventive activity by giving other inventors ideas. As we saw in chapter 7, a given patent protects an inventor only in the country in which he has taken it out. For example, a U.S. patent will offer protection in the United States, but in no other country. If the inventor seeks patents elsewhere, he must obtain them separately in each country where he desires protection. Unfortunately for him, it is not simply a matter of taking his original patent application and sending it to other countries. Each country has its own unique patent system with its own special requirements, so that each application must be tailor made. As was discussed earlier patent systems vary among countries on such things as filing fees, degree of novelty required for the invention, duration of protection, thoroughness of the patent search, technical fields covered, and working requirements (that is, requirements that the invention be commercialized after a specified period of time).

To the extent that an international patent system exists, it is governed by the Paris Convention for the Protection of Industrial Property. The Paris Convention, which originated in 1883, has since undergone a number of revisions. Today it has approximately ninety national signatories. The Paris Convention allows each country to establish and maintain its own patent system, subject to very few restrictions.[24] The most significant restrictions are contained in Articles 2 and 4 of the convention. Article 2 deals with what is called the "National Treatment for Nationals of Countries of the Union." It states, in part: "Nationals of any country of the Union shall, as regards the protection of industrial property, enjoy in all the other countries of the Union the advantages that their respective laws now grant . . . to nationals; all without prejudice to the rights especially provided for in this Convention."[25] In other words, each national signatory to the convention must extend to foreign patent holders the same rights and privileges extended to its own citizens on patent matters.

Article 4 deals with what is termed the "Right of Priority." If someone files a patent application in one of the Union countries, he is allowed up to twelve months from that date to file an application in other Union countries. The effective filing date for all the applications is that of the first filing, which may be important in determining an invention's priority over competing inventions that came into being at the same time.

It is this international patent system, as defined by the Paris Convention, that the LDCs find objectionable. They claim that rather than stimulate invention, the system suppresses it in the Third World. To support their view they point out that the great bulk of patents held in the Third World countries are foreign owned (approximately 85 percent, according to one source),[26] and that very few of these foreign-owned patents are actually worked—that is, put into production—in the patent-granting LDC (approximately 5 percent).[27] Consequently, the patent systems in the Third World primarily benefit foreign businesses and foreign inventive activity.

The statistics that demonstrate the small level of domestic patenting in LDCs cause us to step back and reflect upon the following question: Do the patent systems in LDCs serve the functions for which they are designed? That is, (1) Does the industrial property protection they offer encourage the introduction of innovations into the economy?; and (2) Does the public disclosure of inventions stimulate further innovative activity? In regard to the first question, the statistics previously cited clearly suggest that very few patented inventions are actually worked in LDCs either by the patent holder or through a license to a local producer. Hence, existing patent systems do not appear to encourage the introduction of innovations in LDCs. It would seem that this problem could easily be remedied by including in the LDC patent law a requirement that patent holders work their inventions within a certain period. If the patents are not worked, the law could require compulsory licensing to a domestic operation or revocation of the patent. The problem with taking this approach is that while the Paris Convention allows compulsory licensing, it does so under conditions that may entail the passage of substantial, and possibly crucial, amounts of time.[28] The convention stipulates that ''A compulsory license may not be applied for on the ground of failure to work or insufficient working before the expiration of a period of four years from the date of filing of the patent application or three years from the date of the grant of the patent, whichever period expires last. . . .''[29] Since it may take two or three years for a patent to be filed and granted, many years could easily pass before the compulsory licensing requirement could take effect.

As to revoking the patent because it has not been worked, the convention states that ''No proceedings for the forfeiture or revocation of a patent may be instituted before the expiration of two years from the grant of the first compulsory license.''[30] This adds at least another couple of years to the process of introducing an innovation into the economy in the case where the patent holder refuses to work his invention in an LDC.

In regard to the second question previously raised, the disclosure of an invention in an LDC will probably not stimulate further innovative activity in that country simply because it lacks the scientific and technological capability to transform disclosed inventions into new ones.

By adhering to the Paris Convention, LDCs apparently do not realize the most fundamental benefits that patents are intended to provide—that is, the introduction of inventions into the economy and the stimulation of further invention through disclosure. Problems with the international patent system go far beyond this, however. For one thing, they have substantial negative economic implications. Foreign holders can use patents to give them a monopoly position that boosts the price of their products, which contributes to inflationary pressures in LDCs. Furthermore, if these products are imported items, the LDC's balance of payments position will suffer. Foreign patent holders can also prevent domestic production of a patented good in an LDC, having obvious technology transfer and employment implications. Foreign patent holders can also use their position to keep other foreign competitors out of an LDC, even if they themselves neither produce a good in the country nor import it. In short, it is possible for an LDC to lose substantial control over sectors of its economy if it follows Paris Convention rules.

Other problems with the international patent system are rooted in the LDC's lack of administrative and technical competence. The proper maintenance of a worthwhile patent system requires a substantial administrative apparatus and the employment of competent scientists and engineers as patent examiners. However, these are lacking in most LDCs, and as a result the granting of an LDC patent can be a haphazard event. One negative consequence of this is that undeserving inventions that are neither unique nor original, and possibly even fraudulent, may be given patent protection, creating unnecessary monopolies. The administrative problems could be overcome to a degree if LDC patent awards could be made contingent upon the granting of a patent in a developed country with good examination capabilities. However, once again a possible solution to an LDC problem is thwarted by the Paris Convention, which maintains that patents granted in two or more distinct countries are independent of each other " . . . as regards the grounds for nullity and forfeiture. . . ."[31] That is, one country cannot deny an inventor a patent simply because another country does so.

Of course, LDCs are not compelled to adhere to the Paris Convention. They can devise intellectual property laws that are well suited to their needs even if they are in violation of the principles set out in the Paris Convention. They can follow the examples of India and Mexico, who in 1972 and 1976 respectively put into effect new patent laws that were basically in compliance with the Paris Convention, but which included articles stating that compulsory licensing may be required any time the government determines that this would be necessary to serve the public interest.[32]

The problem with such an approach is that it may discourage foreign investment activity in the country. One proponent of the existing international patenting system has pointed out that between 1970 and 1976, the number of foreign patent applications in India decreased by half, while domestic patent applications did not increase.[33] Is this bad for India? Not necessarily. While foreign patenting clearly declined after the introduction of its new patent law, it is difficult to say just what this means. Nonetheless, the unknowns associated with the decrease in foreign patenting are unsettling.

Rather than have each country set out on its own to deal with patent problems, the Third World would prefer to have the patent system overhauled at the international level. There are a number of changes that LDCs would like to see in the system. For one thing, they would like the Paris Convention to allow them certain concessions. For example, they want to be able to institute two-tier national patent systems. Full protection would be offered to nationals in order to stimulate domestic inventiveness, while more limited protection would be offered to foreign patent holders, entailing more stringent working requirements.

LDCs would also like to see changes made in the Paris Convention that would ease the burden of effectively administering a national patent system.

In particular, they want the convention to allow them to grant or deny patents on the basis of examinations conducted in the advanced countries.

Some of the criticism of the international patent system is directed specifically at the organization responsible for administering the Paris Convention: the World Intellectual Property Organization. LDC critics believe that WIPO is more concerned with maintaining a tradition of protecting inventors through patent law than with promoting the public interest. C.V. Vaitsos has pointed out that the model code developed by WIPO for adoption by Third World countries is even stronger in upholding the traditional values than the Paris Convention, for all its shortcomings, requires.[34]

A group of experts has been working under UNCTAD auspices on a revision of the international patent system since 1974, while WIPO has been engaged in similar efforts since 1975 through an "Ad Hoc Group of Governmental Experts on the Revision of the Paris Convention." The work of these two groups was to culminate in a Diplomatic Conference for the Revision of the Paris Convention. Such a conference was held from 4 February-4 March 1980 in Geneva. Unfortunately, the conference could reach no agreement on adopting rules of procedure for revising the Paris Convention. Even this limited progress was threatened by patent authorities in the West who felt that the procedural rules were ramrodded through by the LDCs.

A second session to the Diplomatic Conference was held in Nairobi, Kenya, from 28 September-24 October 1981 where discussions went beyond procedural issues. Attention focused on the inclusion of provisions into the Paris Convention for forfeiture of nonworked patents as well as the justification for compulsory licensing of patents used abusively. LDCs want to make forfeiture and compulsory licensing easier to undertake in order to discourage MNCs from holding patents without working them. The Western countries were split on these matters, with Canada and some of the smaller European countries sympathetic to the LDC position, the United States staunchly against it, and the other Western countries occupying a middle ground. It became clear at this second session that the Third World was making progress in having some of its views accepted.

Codes of Conduct for Technology Transfer

One of the LDCs' bitterest complaints against MNCs is that they use technology transfer to exploit the Third World. Consequently, since the early 1970s the LDCs have been working to establish international rules, so-called codes of conduct, to govern technology transfer.

The attempt to establish guidelines prescribing international business behavior long predates the LDC efforts of the 1970s. For example, in 1949 the International Chamber of Commerce (ICC) drafted a "Code of Fair Treatment for Foreign Investments."[35] This code was not adopted by any countries at the time of its formulation, although an updated version was adopted by the ICC in 1972. Interestingly, the original ICC code was designed to protect international business enterprises from the arbitrary actions of states. The codes formulated two and three decades later were, in contrast, concerned with protecting states from the arbitrary actions of large international corporations.

In the 1970s, codes of conduct emerged from various quarters. Following the lead of Caterpillar Tractor Company in the early 1970s, a number of the most prominent MNCs put together short statements of business principles, which described their ethical responsibilities. Included in this list of MNCs were such giants as Eastman Kodak, International Harvester, Scott Paper Co., Alcoa, and Gulf Corporation. There were two motivating forces behind the behavior of these MNCs. One was the perceived need to answer criticisms by the LDCs that they acted irresponsibly in their dealings with the Third World. The other was the recognition that a series of scandals involving pay-offs and other forms of corruption among a number of prominent American MNCs in the early 1970s required a major public relations effort to assure the American public of the good intentions of the MNCs.

The central focus of corporate codes of conduct was on such things as pricing, competitive behavior, payment of commissions, and standards of product quality. To the extent that the codes dealt with the issue of technology transfer, corporations would typically state that their technology was used to benefit people throughout the world in a number of ways: (1) the products produced by the MNCs improved living conditions; (2) people abroad received technical training for using and/or manufacturing the products; and (3) licenses were available on reasonable terms for production of goods by enterprises abroad. From the point of view of the Third World, statements such as these were self-serving, insubstantial exercizes in public relations and not much more.

Western governments also assumed some initiative in issuing a code of conduct for international business behavior. This was done under the auspices of the Organization of Economic Cooperation and Development (OECD) and ultimately was embodied in a document called "Guidelines for Multinational Enterprises," which was promulgated in 1976 and revised in 1979.[36]

The OECD Guidelines address a broad range of issues, including technology transfer. In a section titled "Employment and Industrial Relations," Item 5 states that MNCs should attempt to upgrade the work skills of local employees through training.[37] Such training, of course, has a strong technical element to it. In a section entitled "Science and Technology," the guidelines

identify three areas of responsibility to be assumed by multinational enterprises. That is, they should: (1) attempt to see to it that their activities contribute to the building of national scientific and technological capabilities in
the host country; (2) adopt to as great an extent as possible practices that
will increase the diffusion of technology; and (3) grant licenses on reasonable terms and conditions.[38]

An important thing to note about the OECD Guidelines is that they are
precisely what they are titled—guidelines. They *suggest* behavior that multinational enterprises should follow. They are *not compulsory*.

The countries of the Andean Common Market (ANCOM) took a far
more hard-nosed approach in dealing with multinationals.[39] (Originally,
these countries included Bolivia, Chile, Colombia, Ecuador, and Peru.) The
ANCOM investment provisions were viewed as extremely harsh. As foreign
investment in these countries declined, and their economies stagnated, it
became clear to many observers that the strict ANCOM investment provisions were self-defeating.[40]

On the international level, the Commission on Transnational Corporations was established by the United Nations in 1974, and one of the chief
functions of the Commission was the establishment of a code of conduct
governing the behavior of transnational corporations. The evolving UN
code of conduct dealt with a great many issues, including the transfer of
technology. However, in the mid-1970s, UNCTAD began focusing on technology transfer by the multinationals, and the Commission on Transnational Corporations deferred to UNCTAD in regard to those aspects of a
code of conduct that dealt specifically with technology transfer.

UNCTAD's Code of Conduct

UNCTAD's attempt to develop a code of conduct for technology transfer
dwarfs all others. The origins of this code are rooted in the Pugwash Conference held in Switzerland in April 1974. At this conference, a small number
of experts developed a draft version of a code of conduct governing the
transfer of technology. Because the draft focused entirely on the responsibilities and obligations of MNCs, and avoided dealing with the responsibilities and obligations of host governments, it clearly would appeal to
LDCs and be viewed with skepticism by multinationals. The Pugwash draft
was promoted by the Group of 77 when, at the behest of the chairman of
the Group of 77, it was published and circulated by UNCTAD on 15 July
1974.[41] It was effectively adopted by the Group of 77 when, at an UNCTAD
meeting in May 1975, they submitted for consideration a somewhat revised
version of the document as representative of their views on a code of conduct for technology transfer. Ultimately, the market-economy countries

(called group B countries in UN terminology) and socialist countries group D) would submit their own versions of a code of conduct.[42] At the UNCTAD IV meeting in Nairobi in 1976 it was agreed that an intergovernmental group of experts be created to prepare a draft code. Thus, by 1976 the development of a code of conduct of technology transfer was official UNCTAD business.

The gap separating the market economies from Third World countries was very apparent in the negotiations on the code. The gap was not merely the consequence of two parties staking out negotiating positions at opposite ends of a continuum. Rather, it reflected the fact that the North and South were approaching the issue from two entirely different conceptual platforms. For example, in the Third World, governments represent domestic purchasers of technology and negotiate on their behalf. In market economies, governments do not generally involve themselves in such negotiations. In fact, the freer the market economy, the weaker the connection between government and private enterprise. As a consequence of this fact, while Third World representatives can negotiate on behalf of technology purchasers, market-economy representatives do not negotiate on behalf of private enterprises. Third World representatives do not go along with this market-economy position and view it as ploy to avoid reaching a significant agreement. One prominent Third World observer has pointed out that when it is convenient, market-economy governments have no difficulty regulating the economic activities of their private enterprises, as when they participate in GATT agreements, impose export restrictions, and negotiate mutually acceptable tariffs on imports.[43]

For their part, many representatives of the market economies see the positions taken by the LDCs as unrealistic and unreasonable. One American observer has cynically suggested that an important reason for the lack of agreement at negotiating sessions on the code of conduct is that the Third World experts do not have the technical competence to make worthwhile contributions to the sessions.[44] He observes that these so-called experts are government workers who are in effect professional conference-attenders, going from one conference to another.

The single greatest unresolved issue between North and South in the negotiations is the matter of the legal character and effect of the code. Group of 77 negotiators want the code to be compulsory. In their view, without a compulsory agreement the code would be largely meaningless. For their part, negotiators from the market economies want the code to assume the character of a guideline. They are not ideologically disposed to overseeing and imposing technology-transfer regulations on private enterprises. Furthermore, there is some question as to whether such regulation is compatible with domestic law.

Restrictive Practices in Technology Transfer

Substantively, the heart of the draft code of conduct is chapter 4, which identifies twenty "restrictive practices" in technology transfer.[45] All parties in the negotiations (G-77, group B and group D) have reached substantial agreement on most of the items to be included in this part of the code. In those areas where there are phrasing problems, we generally find the market-economy countries attempting to tone down the wording of the restrictions and the G-77 countries attempting to strengthen it. The Socialist countries typically occupy a middle position.

The practices covered in this section of the draft code bear on the licensing of technology and trademarks and were discussed in chapter 7. The degree to which the practices are in fact restrictive depends upon one's perspective. Some of them are clearly restrictive and noncompetitive, as in the case of requirements made by a technology donor limiting the amount of research that can be undertaken by a technology recipient. Others are questionable. For example, the G-77 countries would like to include in the list of restrictive practices the use of quality-control requirements by technology donors. Clearly, the imposition of quality control requirements by the donor may be legitimate when such requirements are designed to assure good production practices. On the other hand, technology donors can also use quality control provisions as a leveraging tool that gives them significant control over the activities of technology recipient.

Stalemate on the UNCTAD Code of Conduct

After a great flurry of activity in the mid-1970s, little progress was made in putting together a code that was acceptable to all parties. It became clear that the market economies would never agree to a code that was compulsory. By the same token, it was clear that the Group of 77 viewed a voluntary code as having little or no value. Neither side would give in on this matter. Furthermore, there was great dissatisfaction among the market-economy countries with the one-sided nature of the draft that was evolving. Clearly, the draft code was directed against multinational corporations. Virtually all the restrictions in the code applied to them. Host governments had few responsibilities defined in the code. As of the early 1980s, it was questionable whether anything concrete would emerge from the exercise of the previous decade.

Other Demands

The resolution of Third World demands relating to the law of the seas, the international patent system, and codes of conduct governing technology transfer will never be clear cut. To the extent that specific solutions to the issues are achieved, they will be temporary because of constant changes in the technological, political, and economic arenas. What is clear is that the issues will remain with us for a long time in one form or another.

The three issues discussed in detail earlier constitute some of the more significant demands made by the Third World in regard to improved access to the fruits of Northern technology. By no means do they exhaust the list of demands. International meetings are replete with LDC demands, some major, some very trivial. Such demands have come to be expected at these meetings as part of the Third World effort to usher in a new international economic order. In order to give a fuller flavor of the range of demands, three additional areas that have led to North-South confrontations will be briefly discussed. They focus on: the World Administrative Radio Conference, the UN Conference on Science and Technology for Development, and the Moon Treaty.

World Administrative Radio Conference

Much of the electronic communications in our world utilize what is called the radio spectrum. (Other communications occur over wires or via fiber optics.) The radio spectrum is composed of a continuous range of radio frequencies employed for such purposes as AM and FM broadcasts, citizen's bands, VHF and UFH television, microwave relay, radar, and navigation. The basic unit of frequency measurement is the Hertz (Hz), which represents one cycle per second. The radio spectrum is made up of frequencies ranging from .003 MHz (megahertz) to 300,000 MHz (mega = million). It is divided into eight broad classifications, some of which are familiar to the typical consumer (for example, Very High Frequencies [VHF], 30-300 MHz; Ultra High Frequencies [UHF], 300-3,000 MHz. Both of these frequencies are used for television broadcasts.).

Use of radio frequencies must be controlled carefully, since it is easy for two users utilizing the same frequencies to interfere with each other. Countries deal with domestic use of radio frequencies through domestic regulatory agencies. In the United States, for example, commercial use of the radio spectrum is governed by the Federal Communications Commission (FCC), while military use of the spectrum is controlled by the National Telecommunication and Information Administration (NTIA). Internationally, such regulation occurs through the International Telecommunications Union (ITU), specifically through its subsidiary World Administrative Radio Conference (WARC).

A full-scale WARC meeting is held once every twenty years in order to resolve major issues in the telecommunications area. Most recently, such a meeting convened in 1979. In between these plenipotentiary meetings a number of specialized meetings are held. These deal with specific technical issues that need resolution. Regional meetings are also occasionally held to discuss outstanding issues bearing on a given WARC region.

International regulation of the allocation of radio frequencies (allocation both to countries as well as to different uses) has reached the point where it can have enormous economic and military implications for all countries. The problem is that the most desirable ranges of the radio spectrum are being rapidly depleted. These are the ranges where existing technology allows relatively cheap use of radio frequencies. At the very highest end of the radio spectrum, there is room for greater use. However, the technology for the cost-effective use of these extra high frequencies does not exist.

In surveying this situation, LDCs have become alarmed. They fear that the current first come, first serve approach to allocating radio frequencies will mean that all that will be left when they are ready to utilize the spectrum more heavily will be the undesirable frequencies. Their concern focuses particularly upon use of satellites for communications purposes. The satellites that they need for telecommunications are located in geostationary orbits some 22,300 miles over the earth's equator. Unfortunately, there are technical limitations on the number of geostationary telecommunications satellites that can be put in orbit. Third World countries see that geostationary satellite slots for their use are scarce and growing scarcer all the time.

Interestingly, the principal use that LDCs will have for satellite communications is domestic communications, not international communications. Unlike the developed countries, many LDCs do not have elaborate terrestrial communications networks already in place in their countries. By utilizing satellites for telecommunications purposes, they will be able to avoid the expense of building costly terrestrial systems.

Third World concerns about being squeezed out of the international telecommunications system were growing rapidly by the time the 1979 WARC was held. In preparation for the conference, G-77 countries met together to develop a common position. The developed countries viewed the G-77 moves with great anxiety, since they feared that the Third World countries would turn WARC 1979 into a forum for making unreasonable political demands. Traditionally, WARC conferences have been low-key technical affairs that, because of their non-political nature, have contributed substantially to the nurturing of an effective global telecommunications system. However, in view of the behavior of LDCs in other technical conferences, and in view of the noises they were making in preparation for WARC 1979, it seemed likely that the conference would be wracked by political discord. WARC 1979 began inauspiciously, when the Third World countries—comprising more than 100 of the 154 delegates—made it clear that they wanted the conference chairman to come from a non-aligned country. Ultimately, a chairman was chosen from Argentina, and a vice-chairman from New Zealand.

Despite these inauspicious beginnings, WARC was not converted into a political three-ring circus. Hard bargaining was undertaken by all parties. Everyone gained a little, and everyone lost a little. The most contentious issue involved Third World demands that a frequency allocation system be created that would reserve in advance frequencies for LDC use, even before the LDCs actually had the capacity to use them. The Third World countries hoped that such a system of allocation would assure that the radio spectrum would not be gobbled up before they had a chance to have their fair share of good frequency allocations. The United States was strongly opposed to such a system, preferring to continue with the present first come, first serve system of frequency allocation. Rather than stall the conference on this one issue, the delegates agreed to take it up at a later specialized WARC to be held in the mid-1980s.

UN Conference on Science and Technology
for Development

The UN Conference on Science and Technology for Development was held in Vienna late in the summer of 1979. The convening of the conference entailed several years of preparations, both in the United Nations and in the 141 countries that sent representatives to Vienna. One estimate put the cost of the conference at fifty million dollars.[46] By all accounts the conference was a major undertaking.

It was hoped that such a conference would lead to the mobilization of S&T for the purpose of helping Third World countries to develop economically. As was mentioned earlier in this chapter, initial awareness of the importance of S&T did not really emerge until the early 1960s, and this was reflected in the UN Conference on the Application of Science and Technology for the Benefit of the Less Developed Areas, as well as the establishment of bureaucratic machinery in the UN system to deal with the issue of science, technology, and development.

Unfortunately, the Vienna conference did not meet the expectations of the optimists. One problem was the sheer cumbersomeness of the proceedings, which catered to some 141 diplomatic delegations and a total of well over two thousand individual participants. It is difficult to obtain consensus among such an unwieldly group. Another problem was the extreme breadth of the agenda, which dealt with all manner of topics on science, technology, and development, as well as other topics unrelated to S&T. Still another problem was the inherent conflict in the positions staked out by the North and South, where, as frequently happens in such encounters, strong Group of 77 demands were made for better access to Northern technology, and Northern responses emphasized the impossibility of meeting such demands. Many of the G-77 demands were only peripherally related to the subject matter of the conference (for example, demands regarding transfer pricing of MNCs, repatriation of profits, local materials content requirements), and were in fact being handled much more effectively in other forums.

Two tangible outcomes of the conference were the agreement to create by means of voluntary contributions a \$250-million Fund for Science and Technology in 1980-1981, and the agreement to restructure the UN's S&T bureaucracy. Originally, the LDCs demanded that the S&T fund be much larger, with an annual budget of two billion dollars by 1985 and four billion dollars by 1990 (these funds were to be raised by a tax on the developed countries). Within a few months after the conclusion of the conference it became apparent, however, that even the \$250-million target would not be reached, as actual pledges to the fund were very low.

Perhaps the single most significant outcome of the conference was intangible. That is, national preparations for the conference as well as the conference itself, increased, as never before, first-hand awareness, in both the developed and developing countries, of the importance of S&T to the development effort. In the United States enthusiasm for S&T as development tools was so great that the Carter administration proposed the creation in the United States of an Institute for Scientific and Technological Cooperation (ISTC). ISTC's initial budget would be a substantial ninety million dollars. However, once the UNCSTD fanfare was over, delegates returned home, and business-as-usual attitudes prevailed, American enthusiasm for S&T as development tools waned, and ISTC never received congressional approval.

The Moon Treaty

The full name of what has come to be known as the Moon Treaty is "Agreement Governing the Activities of States on the Moon and other Celestial Bodies." The text of the treaty originated in the Committee on Peaceful Uses of Outerspace. It was adopted by the UN General Assembly without a vote or opposition on 5 December 1979.[47] For the pact to become law, it must be ratified by at least five countries, although it will only apply to the signatories.

There are a number of parallels between the Moon Treaty and the UN Conference on the Law of the Seas. For example, the Moon Treaty declares the moon and its resources to be the common heritage of mankind; and it calls for the establishment of an international authority to govern the exploitation of the moon's resources.[48] The similarities between the Moon Treaty and the provisions debated at UNCLOS III are not totally coincidental, inasmuch as the UNCLOS III and Moon Treaty proceedings occurred concurrently.

While UNCLOS III attempted to deal with specific issues (for example, regarding the functioning of an International Seabed Authority), the Moon Treaty is basically a statement of general principles lacking specificity. It can be viewed as the first step in a two step process. Once the general principles have been accepted by the world community, the next step will be to work out the specific details.[49]

The circumstances surrounding the Moon Treaty point up dramatically the glaring disparities between the haves and have-nots of the world. Nowhere is the contrast between haves and have-nots greater than in the case of the exploration of outer space. The level of S&T involved, coupled with the enormous commitment of resources required, puts space travel out of the reach of all but a handful of countries for generations to come. To date, only one country has demonstrated an ability to land men on the moon (the United States). Yet the Third World countries nonetheless demand their fair share of the fruits of outer-space activities undertaken by one or two countries which have developed all the technology, assumed all the risks, and raised all the needed finances.

Within several months after the General Assembly adoption of the Moon Treaty, it became apparent that the treaty would have serious trouble being ratified by the U.S. Senate. U.S. corporations with an interest in outer-space activities publicly declared that the treaty was an attempt by Third World countries, in collusion with the Eastern bloc, to undermine the free-enterprise system in the industrial West.[50] Groups of concerned citizens caught up with the idea of space exploration also organized lobbying efforts to defeat ratification of the treaty.[51] The great fear in the United States was that the existence of restrictive international legislation governing activities in outer-space would lessen the incentive both in government and the private sector to explore outer space.

Conclusions

A careful review of the North-South conflict makes it evident that there are no villains and no heroes in this story. Each side makes legitimate claims. At the same time, each side appears somewhat blind to the claims of the other. A fair assessment of the global economic system today shows that it does indeed often work against the interests of the Third World. By the same token, an examination of what it is that motivates MNCs to take enormous risks and raise large quantities of capital suggests that if many of the Third

World demands are met, much of the incentive for future major technological advances will be gone.

There is no simple explanation of the sources of conflict between North and South. Clearly, a number of factors are at play. Several of the more significant factors are:

Economic. There is an enormous economic gap between the developed and underdeveloped countries, and it does not appear to be closing. One major cause for this gap is the existence of large technological capabilities in the North and their absence in the South. The obvious disparity between the rich and poor in the world is a sore point both with the Third World, which would like to see a redistribution of wealth, as well as with the developed free-market countries, which fear that such a redistribution would lower their standards of living.

Ideological. The wealthy countries of the industrialized West credit their successes in large measure to free enterprise. They firmly believe that the economic incentives and competition of the unfettered market place keep them innovative, efficient, and prosperous. The poor countries of the South believe that the free-market model of the industrialized West does not apply to them, and that for economic development to take place, they must engage heavily in central planning. Each side displays a great deal of antipathy toward the approach taken by the other. To free marketers in the North, the central planning approaches of the South smack of a welfare mentality, reflected in international meetings by constant LDC demands requesting something for nothing. To the socialist-oriented Third Worlders, the free-market approach is simply a ploy used by MNCs to cover up their depredations of the economically weak.

Cultural. The cultural chasm separating North from South is enormous. An enumeration of the implications of these differences could fill a volume. Only one manifestation of the cultural gap will be offered here. That is, we find that North and South often take dramatically different approaches to problem solving. For example, when North and South come together in the various forums we have discussed in this chapter, we find the North sending large teams of technically qualified people to conferences, armed with reams of data, and looking for technical solutions to problems that often have a strong social character to them. In contrast, representatives from the South at these conferences frequently are technically deficient and focus their attention on the broader political issues underlying problems. In taking this approach, they may well overlook the technical dimensions of the problem. Unfortunately, when the two groups get together at conferences, misunderstandings frequently result. To Northern participants, Third World representatives are technically deficient and make broad, sweeping, political motions that lack substance and are irrelevant to the issues at hand. To Third World participants, the technical tunnel vision of Northern representatives is at best viewed as a poor approach to problem solving, and at worst seen as a cover for some hidden sinister objectives of MNCs and other agents of imperialism.

Historical. Certainly, historical factors play a significant role in contributing to conflict in North-South relations. It is difficult for representatives of the Third World to forget that a number of the most powerful and technologically advanced countries of the North were their colonial masters not so long ago. There is clearly a concern among Third Worlders today that the brutal political and economic domination of the past not be supplanted with a more subtle technological domination.

What are the prospects for North and South getting together and resolving their differences? Unfortunately, so long as the Group of 77 continues with its confrontation approach to discussing these issues, it does not seem likely that many differences will be resolved. G-77 representatives may feel great visceral satisfaction when they are able to muster absolute majorities at conferences to support Third World views and are able to issue all sorts of regulations that are unacceptable to the North. But they have to recognize that the momentary psychic gratification they experience is no substitute for the concrete results they really want. When they take a confrontation approach to resolving issues with the North, they generally find that they come away from these encounters empty-handed, except, perhaps, for a piece of paper stating some resolution they were able to ram through the proceedings.

Furthermore, the Northern countries are able to walk away from these encounters with clear consciences, since it is obvious to them that the absurdity of G-77 demands shows that the Third World is not really serious about finding solutions to the problems it identifies.

The Third World should ease its confrontation approach for a very simple reason: it is not getting LDCs anywhere. Looking back to the early 1970s, it can be seen that in the intervening years the Third World has achieved very little as a consequence of its aggressive style. A workable law of the seas treaty has not come into being; the international patent system remains basically unchanged; a universal code of conduct is unlikely to be accepted; very little funding is available for the promotion of science and technology in LDCs; and so forth.

Perhaps Third World countries should take note of actors in their ranks who made spectacular economic advances in recent years without resorting to confrontational excesses, actors such as Singapore, South Korea, Hong Kong, and Taiwan. Perhaps a study of the so-called four little dragons would show that a country's energies are more effectively employed improving an economy through internal efforts than trying to right the wrongs perceived to be perpetrated by the rich countries of the North.

Notes

1. *General Agreement on Tariffs and Trade*. Preamble.

2. United Nations, Conference on Trade and Development I, "Joint Declaration of the Seventy-Seven Developing Countries," *Proceedings*, vol. 1 (1964), pp. 67-68.

3. United Nations, Sixth Special Session of the General Assembly, *Resolution 3201 (S-VI)* and *Resolution 3202 (S-VI)*, May 1974.

4. E.B. Haas, M.P. Williams, and D. Babai, *Scientists and World Order* (Berkeley: University of California Press, 1977), p. 247.

5. Advisory Committee on the Application of Science and Technology, *World Plan of Action for the Application of Science and Technology to Development* (New York: United Nations, 1971).

6. United Nations Economic and Social Council, *Report of the UN Water Conference*, Mar del Plata, 14-25 March 1977, p. 82.

7. United Nations Economic and Social Council, *Report on the UN Conference on Desertification*, Nairobi, 29 August-9 September 1977, p. 87.

8. Ibid., p. 93.

9. Elliot L. Richardson, "Power, Mobility and the Law of the Sea," *Foreign Affairs* 58 (Spring 1980):904.

10. J.L. Mero, *The Mineral Resources of the Sea* (New York: Elsevier, 1965).

11. Quoted in V.E. McKelvey, "Seabed Minerals and the Law of the Sea," *Science* 209 (15 July 1980):464.

12. United Nations, General Assembly, (A/C.1/PV 15/5), November 1, 1967. United Nations, General Assembly, (A/C.1/PV 15/6), November 1, 1967.

13. United Nations, General Assembly, *Declaration of Principles Governing the Seabed and the Ocean Floor, and the Subsoil Thereof, beyond the Limits of National Jurisdiction, Resolution No. 2749* (XXV), 17 December 1970.

14. S. Brown, N.W. Cornell, L.L. Fabian, and E.B. Weiss, *Regimes for the Ocean, Outer Space, and Weather* (Washington, D.C.: The Brookings Institution, 1977), p. 74.

15. R. Hudson, "For a Law of the Seas—Forecast: Storms Ahead," *Current* No. 198 (December 1977):46.

16. P.L. Dalkiewicz, "Ocean Exploration: Risks and Prospects," 5 February 1981. Unpublished manuscript.

17. C.G. Welling, "The Ocean's Waiting Mineral Resources," *Stockton's Port Soundings* 3 (August 1980):8.

18. McKelvey, "Minerals and the Law," p. 465.

19. "Developing Countries, USSR, US Propose Differing Systems for Exploiting Seabeds," *UN Monthly Chronicle* 13 (October 1976):37-38.

20. Ibid., p. 38.

21. Ibid., p. 39.

22. These issues included: (1) establishing rules to cover the areas of the sea where there is overlapping jurisdiction of neighboring states over the EEZ and the continental shelf; (2) participation in the convention by various nongovernmental groups, such as the Palestine Liberation Organization; (3) the establishment of a Preparatory Commission that would put into effect the machinery that would lead to the functioning of the International Seabed Authority; and (4) the inclusion of a grandfather clause into the convention that would allow companies to begin mining the seabeds prior to the convention's ratification. "Seabed Law Conference Adjourns Until August, No Accord on Remaining Issues," *UN Chronicle* 18 (June 1981):19.

23. See, for example, M.T. Lilla, "Third World's Sea Pact Takes U.S. for a Ride," *Wall Street Journal*, January 26, 1981, p. A20.

24. *Paris Convention for the Protection of Industrial Property*, Revised in Stockholm, 1967.

25. Ibid., Article 2, Paragraph (1).

26. H.P. Kunz-Hallstein, "The Revision of the International System of Patent Protection in the Interest of Developing Countries," *IIC: International Review of Industrial Property and Copyright Law* 10 (1979):658.

27. Ibid.

28. *Paris Convention*, Article 5, Section A, Paragraph (2).

29. Ibid., Article 5, Section A, Paragraph (4).

30. Ibid., Article 5, Section A, Paragraph (3).

31. Ibid., Article 4bis, Paragraph 2.

32. *The Patents Act* (India), 1970, Article 97; *Law on Inventions and Marks* (Mexico), 30 December 1975, Articles 3 and 56.

33. Kunz-Hallstein, *Industrial Property and Copyright Law*, p. 660.

34. C.V. Vaitsos, "The Revision of the International Patent System: Legal Considerations for a Third World Position," *World Development* 4 (1976):85-102.

35. R. Black, S. Blank, and E.C. Hanson, *Multinationals in Contention* (New York: The Conference Board, 1978), p. 135.

36. Found in OECD, *International Investment and Multinational Enterprise* (Paris: Organization for Economic Cooperation and Development, 1979).

37. Ibid., p. 20.

38. Ibid., p. 21.

39. Andean Common Market, *Common Regime of Treatment of Foreign Capital and Trademarks, Patents, Licenses, and Royalties* (1970).

40. See, for example, "South America: Why the Andean Pact is Falling Apart," *Business Week*, 21 September 1981, p. 46.

41. United Nations, Conference on Trade and Development, *The Possibility and Feasibility of an International Code of Conduct in the Field of Transfer of Technology*, (TD/B/AC.11/L.12), (1974).

42. For a discussion of the negotiating positions of G-77, group B, and group D countries, see J.W. Skelton, "UNCTAD's Draft Code of Conduct on the Transfer of Technology: A Critique," *Vanderbilt Journal of Transnational Law* 14 (Spring 1981):381-396.

43. M.S. Wionczek, "Code of Conduct on Transfer of Technology— When and Why?)," in M. Finnegan and R. Goldscheider, ed., *The Law and Business of Licensing* (New York: Clark Boardman, 1980), pp. 520.381-520.390.

44. H.O. Blair, "Technology Transfer as an Issue in North/South Negotiations," *Vanderbilt Journal of Transnational Law* 14 (Spring 1981): 301-326.

45. United Nations, Conference on Trade and Development, *UNCTAD Draft International Code of Conduct on the Transfer of Technology* (TD/CODE TOT/25), June 2, 1980.

46. "Talking Technology in Vienna," *The Economist* 25 August 1979, p. 54.

47. UN General Assembly, *Agreement Governing the Activities of States on the Moon and Other Celestial Bodies*, UN Doc. (A/Res/34/68), 1979. Annex to G.A. Res. (34/68).

48. Ibid., Article 11, Paragraphs 1-5.

49. J.H. Works, "The Moon Treaty," *Denver Journal of International Law and Policy* 9 (Summer 1980):281-285.

50. T. Agres, "Moon Treaty Meets Resistance," *Industrial Research and Development* 20 (May 1980):43-44.

51. N. Burnett, "Making Sure We Get Our Share of Space," *Parade*, 31 August 1980, p. 17.

11 Military R&D and Arms Transfers

Introduction

National defense is big business. In recent years, military expenditures have constituted 5-6 percent of global GNP. Roughly two-thirds of a trillion dollars are dedicated worldwide to national defense expenses each year. Most of these expenditures are directed at the support of domestic forces and armaments. However, the figures also include funds that point up the existence of a thriving international arms trade valued at some twenty-five billion dollars per year in the early 1980s.

National defense today is inextricably intertwined with S&T. To a certain extent this has been so since the time many millenia ago when one of our early ancestors first employed a stone axe for the purpose of cracking the skull of another unfortunate ancestor. We have a more sophisticated example of the application of S&T to warfare in the apocryphal story of Archimedes arranging mirrors in such a way that they used sunlight to vaporize a ship in the harbor of Syracuse. Another well-known example is the work of Da Vinci, whose notebooks are filled with drawings and descriptions of many advanced instruments of war.

History is replete with cases of new technologies revolutionizing warfare: The British long bow is said to have brought about the demise of chivalrous combat; the Mongols' use of cavalry introduced a whole new tactical element into combat; the ascendancy of the portable gun, whose ball could penetrate thin metal plate, made armoured suits obsolete. While the nature of warfare has always been determined by the existing technology of war, we find today that S&T are being consciously applied to developing new weapons and military strategies as never before in history.

National defense is clearly heavily dependent upon S&T. By the same token, in a number of the world's most powerful countries, including the United States, the USSR, Great Britain, and France, S&T owe a large debt to national defense. In the United States, for example, roughly one R&D dollar out of four is spent for defense. The figure for the USSR, while not precisely known, is much higher.

In this chapter we examine military technologies both in their domestic environment and in the context of international business transactions. We are concerned not only with a physical description of military technologies and their development, but with their social and economic impact as well.

Military R&D in the United States

It was mentioned in the discussion of science policy in chapter 8, that American scientists and engineers had their first close contact with the military during World War II. During the war, the great bulk of America's scientific and technological talent was mobilized to contribute to the war effort. World War II thus launched the era of big science, where teams of scientists, engineers, and administrators applied themselves to the conscious and systematic resolution of clearly defined problems. This heavy mobilization of scientists and engineers in the United States, as well as in other Allied countries and the Axis powers, resulted in an enormous upgrading of military technology. Among the more dramatic advances resulting from the worldwide employment of scientists and engineers for wartime purposes were the development of operations research, radar, the jet aircraft, the rocket, sonar, and, of course, the atom bomb.

The end of World War II did not bring any long-term relaxation of military tensions. Not long after the conclusion of the war in Europe and the Pacific, the Cold War began. Various Eastern European countries fell under the domination of the Soviet Union, followed by Communist insurgency in Greece and Turkey, the take-over of China by Mao Tse Tung, and the outbreak of the Korean War. While the extraordinary mobilization of America's scientists and engineers ended with the world war, the military continued to maintain close contact with the scientific and engineering community. The Department of Defense (DOD) became entrenched as a major funder of R&D in America.

Military support of U.S. R&D takes several different forms. First, the military directly supports research in its own laboratories (for example, Aberdeen Proving Grounds, Edgewood Arsenal, Naval Research Laboratory, Naval Surface Weapons Laboratory). It also obliquely supports large-scale nuclear weapons laboratories that are managed by universities and funded by the Department of Energy (for example, the laboratories in Los Alamos, New Mexico, and Livermore, California).

Second, the military supports extramural R&D in universities by contracting R&D tasks to them (for example, through the Air Force Office of Scientific Research and the Office of Naval Research).

Third, the lion's share of military R&D support goes to the private sector, where nonprofit firms (for example, Battelle Laboratories, prior to its loss of nonprofit status, Stanford Research Institute, The Mitre Corporation) give advice on design and management questions; and for-profit firms perform research, development, test, evaluation, and management functions, as well as manufacture weapons and delivery systems.

The military R&D budget climbed steadily in the 1950s and leveled off in the 1960s, at which time the budget typically varied between $7 billion and $8.5 billion per year. This leveling of the military R&D budget was off-set, in part, by a dramatic increase in the space budget (with all of its ob-vious military implications) from $400 million in 1960 to $5.1 billion in 1966. But after 1966, the space budget began a steady decline, and this, coupled with the leveling of the defense budget, put serious strains on the allocation of R&D funds for military purposes. One consequence of this was a questioning of the value of military support of basic research, which had no immediate visible pay-offs. A DOD funded effort, called Project Hindsight, found that the contribution of basic research to advances in military technology was practically nonexistent.[1] Of 710 "research events" that ultimately resulted in an operational innovation, only 0.3 percent fell into the category of basic research, 8.7 percent were classified as applied re-search, and the remaining 91.0 percent were categorized as purely tech-nological events. This study, incidentally, was heavily criticized on meth-odological grounds, and the National Science Foundation, whose mission is to support basic research, sponsored a counter study that came up with dia-metrically opposite results. (The NSF study was called Project TRACES.)

The issue of whether DOD should support basic research was resolved in 1970 with the passage by Congress of the Mansfield Amendment, which advised government agencies to support only research that bore directly on their missions. In the defense area, this had the effect of leading to dramatic reductions in DOD support of basic research.

Increases in defense R&D expenditures were modest throughout the 1970s, just roughly keeping pace with inflation. With the election of Ronald Reagan in 1980, a new approach was taken towards defense matters overall. Responding to a perceived weakening of the American defense posture, the Reagan White House earmarked dramatic increases in defense expendi-tures. Military R&D benefitted from this new approach.

Military R&D in Other Countries

The continuous development and deployment of new military technologies is very expensive. Consequently, few countries can afford to follow a course of defense self-sufficiency. No countries approach the massive levels of defense R&D expenditures of the United States and USSR, although defense expenditures are considerable in the United Kingdom and France. Table 11-1 shows the percent of government R&D budgets devoted to defense R&D for the major Western powers and Japan. The table makes clear that defense R&D expenditures consumed far more substantial shares of the overall government R&D budget during the coldest days of the Cold War (that is, during 1961) than during the subsequent thaw. The table also dem-onstrates a much greater commitment to military R&D in the United States, United Kingdom, and France than in West Germany and Japan.

Table 11-1

International Comparison of Percent of Government R&D Budgets Devoted to Defense, 1961-1978

	1961	1967	1972	1976	1978
France	44	35	32	30	33
Japan	4	3	2	2	2
United Kingdom	65	52	43	46	52
United States	71	49	53	51	49
West Germany	22	19	15	12	12

Source: National Science Board, *Science Indicators 1980* (Washington, D.C.: U.S. Government Printing Office, 1981), pp. 214-215.

Defense R&D is more decentralized in the United States than elsewhere. As indicated in the previous section, defense R&D in the United States is performed in all sectors—by government, universities, and industry. R&D objectives are established by a central DOD office for R&D, as well as by the individual services, Army, Navy, and Air Force. Responsibility for an important element of the military arsenal—nuclear weapons—does not lie with DOD, but rather with the Department of Energy. Procurement is largely dependent upon free-market forces: for the most part, any firm can bid on a military project. Contracts are awarded on the basis of cost and performance.

Defense R&D is leaner and generally under greater central control in other countries. For example, the German government designates only two computer companies as qualified to perform R&D and produce goods for the military. The British have nationalized the shipbuilding and aircraft industries, so that much of the military R&D performed in these areas is in fact performed by the government. Sweden's defense budget covers five years, enabling R&D and the introduction of new weapons to be planned rationally.

While Western European countries do not pursue the laissez-faire policies of the United States, they nonetheless exercise less central control over their military R&D than the USSR. Little is known with certainty about military R&D in the Soviet Union. What is known is that military R&D has very high priority. It is generally believed that the bottlenecks encountered in Soviet civilian R&D—for example, shortages of the most rudimentary equipment—are lacking in military R&D. It is also suspected that the relatively high salaries and substantial perquisites offered scientists in the military sector siphon away some solid talent from the civilian sector.

Because the Soviet Union is a Communist state, all R&D undertaken in the country is performed by the government. The formal R&D structure resides in military production ministries. Basic research is undertaken in the institutes of the prestigious Soviet Academy of Sciences as well as in so-called invisible institutes that have no names and are largely unknown to Westerners. These invisible institutes also perform applied research and engage in development work.

Arms, Trade, and the Transfer of Military Technology

Because so few countries are militarily self-sufficient, they must acquire much of their military hardware from other countries. Consequently, the value of international arms deliveries are substantial, hitting roughly twenty-five billion dollars per year in the early 1980s. Data for international arms transfers are provided in table 11-2. The central feature of this table is its dramatic protrayal of the developed countries as arms exporters and the LDCs as arms importers. As the table shows, the dollar value of arms imports is four times as great for LDCs as for developed countries, while the dollar value of LDC exports is less than one twentieth that of the developed countries. We will discuss some of the implications of arms purchases and economic development later in this chapter.

Table 11-2 shows that the global hot spot for arms purchases in recent times is the Middle East. To a large extent, the great political and military instability of the region is a leading cause of the large arms-purchases. In 1979, the Arab-Israeli conflict underlay the purchase of $2.1 billion, $525 million, and $2.0 billion worth of weapons by Iraq, Israel, and Syria

Table 11-2
Value of International Arms Transfers, 1979
(millions of dollars)

	Imports	Exports
All developed countries	4,550	23,725
NATO Europe	2,050	4,150
North America	560	5,750
Warsaw Pace	2,005	11,720
Other Europe	725	900
Oceania	205	30
All LDCs	19,265	1,320
Africa	4,575	150
East Asia	3,145	430
Latin America	1,650	70
Middle East	8,000	510
South Asia	900	15

Source: U.S. Arms Control and Disarmament Agency, *World Military Expenditures and Arms Transfers, 1970-1979* (Washington, D.C., 1982), Table II.

respectively. Further to the east, Iran's purchase of $1.1 billion in weapons (prior to the Shah's ouster) contributed substantially to the flow of arms to the region.

The principal arms exporters are listed in table 11-3. Heading the list is the Soviet Union, which delivered some $10.4 billion in military hardware in 1979. One big-ticket item for the Soviets was surface-to-air missiles (SAMs), some 13, 654 of which were delivered to LDCs between 1977 and 1980.[2] Trailing the Soviet Union in second place is the United States, with $5.6 billion in arms deliveries. The other countries on the list sell substantially smaller dollar amounts of military hardware abroad.

The international sale of arms has grown rapidly in recent years, reflecting in large measure the political instability of the Third World. In the ten years spanning 1970-1979, the value of arms deliveries measured in constant dollars increased by 23 percent. Deliveries to developed countries during this time were relatively level, increasing by only 14 percent in the decade. In contrast, deliveries to LDCs increased by nearly 63 percent.

Motivations for Arms Trade

Motivations for purchasing arms are fairly straightforward. A country buys arms from abroad in order to increase its capacity to engage in armed conflict. In some cases, such as Afghanistan in the late 1970s, governments purchase arms in order to maintain control over domestic elements. These elements may be in open rebellion against the government, or else they may be kept passive through the sheer weight of the government's military prowess. In other cases, arms purchases are made in order to strengthen a country's military position in regard to outsiders. The arms may be used prin-

Table 11-3
Leading Arms Exporters, 1979

	Arms Exports	
	(millions of dollars)	*(percent of all exports)*
Soviet Union	10,400	16.0
United States	5,600	3.1
France	1,400	1.4
United Kingdom	1,000	1.1
West Germany	925	0.5
Czechoslovakia	781	6.4
Italy	550	0.8

Source: U.S. Arms Control and Disarmament Agency, *World Military Expenditures and Arms Transfers, 1970-1979* (Washington, D.C., 1982), Table II.

cipally for defensive purposes (for example, Egypt's stockpiling of advanced weapons) or for offensive purposes (for example, to make possible Libya's invasion of Chad in 1980).

Motivations for arms sales are more complex. Obviously, economics play a role. From the perspective of a firm that manufactures weapons, arms sales abroad mean increased financial returns through expanding markets. They may also mean reductions in the unit cost of producing a good, since economies of scale will be realized through larger production runs. From the perspective of the nation as a whole, arms sales abroad have a positive impact on the balance of trade. In addition, to the extent that they contribute to the arms industry's profitability, they help keep the industry healthy. Finally, since military R&D is generally financed by governments, arms sales may help a government recapture some of its R&D costs.

The economics of arms sales have a negative side to them as well. A country that depends heavily on foreign sales of arms to keep its military industrial base healthy will find the industry increasingly dependent upon these outside sales. This can yield negative economic repercussions if the international demand for the country's products slackens or if the government must cut off certain foreign sales for military or political reasons.

Above and beyond economics, governments may encourage the sale of arms for political purposes, to meet national foreign-policy goals. Through their arms sales, countries hope to maintain some control over the actions of other countries. Arms may be sold to a country as a reward for its friendly posture, as in the case of U.S. sales of advanced arms to Saudi Arabia. They may be sold in order to create a degree of dependence on the supplier, a major objective of Soviet arms sales to Egypt in Nasser's time. They may be sold in order to wean a country from dependence on a foe, as in the case of U.S. arm sales to Egypt after the expulsion of Soviet military advisors. They may be sold in order to maintain a balance of power in a region, an important consideration in U.S. arms sales in the Middle East. They may be sold to a country so that it can perform as a surrogate in conflict situations, as in the case of Soviet arms transfers to Cuba, which sends troops to different areas of the globe on behalf of the Soviets. This listing represents only a sampling of the ways arms sales can be used to further foreign-policy objectives.

The problem with arms sales used for this purpose is that they can backfire badly. For example, arms sales that help maintain a friend in power in opposition to the popular will may lead to the supplier's association with the *ancien régime*. In the event of a coup, the arms supplier may be viewed as an implacable foe by the new leadership. To a certain extent, this explains the hostility of Iranians toward the United States after the Shah's ouster, and the hostility of Nicaraguan centrists and leftists toward the United States for its military support of the Somoza family. Another problem in using arms sales for foreign-policy purposes is that they may result in a tail-wagging-the-dog situation. In these cases, the arms recipient manouevers himself into a position where he calls the shots. Many observers of the United States-Israeli arms supply relationship of the 1970s and 1980s believe this illustrates such a situation.

Finally, arms sales may be motivated principally by military concerns. As in the case of both economic and political motivations, there are many military motivations underlying arms sales. Basically, they can be divided into two categories: arms sales that can have a bearing on potential physical conflicts and transfers directed at on-going conflicts. The military motivation is similar to the political motivation in the sense that it is oriented toward influencing the outcomes of conflict. The chief distinction is one made by von Clausewitz early in the nineteenth century, when he defined war as "politics by other means." That is, military motivations are closely connected with hot conflicts, whereas what we have termed political motivations are oriented toward cool or potential conflicts that have not yet erupted into violent confrontations and are resolved by means short of the application of physical force.

Arms-Transfer Policy

When governments define their policies on the export of arms, they take into consideration their economic, political, and military impacts, and calculate the correspondence of potential policies to national goals. Because in the international arena each country has its own ethos, goals, capabilities, and style of operation, no two countries pursue identical policies on arms sales. Some countries, like individuals, espouse a philosophy of arms control, and while they may produce arms for the defense of the nation, they eschew their large-scale export. Others, particularly small countries intent on producing many of their own weapons, promote arms sales in order to support their local arms industry. Still others, often countries with a pronounced ideological bent, serve as arms clearinghouses, important arms from producers and reexporting them to assist favored causes. And so on.

In this section, we will briefly examine the arms transfer policies of several countries in order to explore some of the variety in these policies.

United States

Since World War II, the United States has been either the first or second largest arms exporter in the world. In the three decades following the war, the nature of U.S. arms transfers has undergone many changes. During the

Cold-War era, arms transfers were largely a product of the global bipolar power configuration, with the United States at one pole and the USSR at the other. U.S. arms transfers were made primarily to Western allies, who came together in 1949 under the aegis of the North Atlantic Treaty Organization (NATO); or to friendly Cold-War outposts, such as South Korea and Taiwan, who were lynchpins in the U.S. policy of containment against Communist expansion. In the late 1940s and throughout the 1950s, the United States was able to exercise substantial control over the terms of its arms transfers, owing to its undisputed leadership in the struggle against a perceived Communist monolith.

In the 1960s, bipolarity began to unravel. In the West, America's allies voiced a desire to play a stronger role in determining Western political, economic, and military policies. The independent course taken by the Europeans was particularly marked in Charles de Gaulle's France, which dropped out of some NATO activities and pursued policies consciously designed to irk the United States. In the East, it became apparent that there were cracks in the Communist monolith when the USSR and China engaged in mutual denunciations and armed conflicts along their lengthy border. Meanwhile, the colonial era was almost completely ended in the 1960s, as dozens of Third World countries gained independence from their colonial masters. While initially the Third World appeared to be a good battleground to act out Cold-War hostilities, by the end of the decade it was evident that Third World countries were more concerned with issues such as anticolonialism, dependence on the West, and access to advanced technology than with the traditional Capitalist versus Communist issues that fed the Cold War. In short, by 1970 the world was a far more complex place than the bipolar world of the 1950s. While the United States and the USSR were the undisputed military giants of the world, their massive arsenals appeared largely irrelevant in a world where conflicts were of a limited nature, and where their resolution was more dependent upon political and economic factors than firepower.

During the 1960s, America's European allies increasingly produced their own weapons. Meanwhile, the emerging Third World countries began shopping around global arms-markets in order to purchase weapons they needed to resolve both domestic and international disputes. As a consequence of these changes, U.S. control over its arms sales began to decrease.

The American arms industry boomed with the Vietnam war. However, the simultaneous fighting of a war on poverty and a war in Asia put severe stress on the American economic system. In order to relieve some of the pressure on the economy, the Nixon Doctrine was put forth in 1972. This doctrine invited America's friends and allies to assume more of the burden for their own defense. It had the effect of increasing their self-sufficiency in producing arms.

As the Vietnam war wound down, the U.S. arms industry went into a slump. Many military contractors decided to abandom military contract work because the boom and bust cycles typical of the arms industry resulted in too much corporate instability. With the decline of the American government as an arms purchaser, foreign military sales increased in importance. For example, between 1970 and 1976, foreign military sales of the twenty-five major U.S. defense contractors increased by 46 percent (in constant dollars), while domestic defense contracts declined in value by 23 percent (in constant dollars).[3] Not only did U.S. arms sales witness a quantitative change during this time, they underwent a qualitative change as well. Prior to the 1970s, U.S. military sales were generally of prior-generation equipment. In the 1970s, foreign demand was for state-of-the-art equipment, and if the United States was to make foreign military sales, it had to release this equipment. Furthermore, some foreign purchasers were more interested in obtaining know-how from the United States for their own arms production than in purchasing the final product, and others conditioned their purchase of American hardware on the willingness of the United Statres to allow them to share in its production.

U.S. arms transfer policy at this time was to support U.S. arms manufacturers in their efforts to sell their wares abroad and to direct the sales toward allies and friendly countries. In some cases, such as in the Middle East, arms transfers were made in a conscious attempt to meet specific U.S. foreign-policy objectives. In others, transfers of hardware were made primarily for economic considerations. With the advent of Jimmy Carter to the presidency in 1977, this policy changed. Carter was committed to slowing down the global arms build-up, so he attempted to restrict U.S. arms sales. In certain instances he refused to sell arms to countries because of concern over human rights violations within their borders. Embassy officials overseas were instructed to refrain from providing U.S. arms salesmen with advice and to avoid helping them make contacts with the local population.

The Carter approach was reversed in 1981 when Ronald Reagan assumed the regins of power. Reagan officials noted that America's unilateral abstinence from arms sales did little to reduce the global arms build-up and that Carter's policy had the effect of hurting the U.S. arms industry. Arms sales should be explicitly used to futher U.S. foreign-policy objectives. Human rights matters, while not to be ignored, would play a secondary role in determining whether or not a given arms sales should be made. Embassy officials were now advised to extend to U.S. arms merchants the same services provided to other businessmen.[4]

Current U.S. arms-transfer policy is not without its drawbacks. First, there is a well-grounded fear that foreign military sales are occasionally made at the expense of U.S. military preparedness. For example, massive shipments of U.S. tanks to Israel in conjunction with the Yom Kippur war of 1973 left the U.S. short-handed. Perhaps even more disturbing is the way sales of advanced hardware to Third World countries ties up American management support capabilities. Third World advanced-arms recipients often lack the needed logistics, training, and technical abilities to support complex weapons systems purchased from the United States and they require the presence of American technical personnel to keep them operating. Yet, there is a shortage of such personnel to meet America's own defense needs. For example, between 1968 and 1978, Air Force Logistics Command personnel decreased from 140,000 to 90,000, while manpower needs to support foreign military sales increased from 2,000 to 4,000.[5]

Second, concern has been expressed that the U.S. defense industry is becoming too dependent on arms sales abroad. From the purview of national defense, the fundamental purpose of a strong defense industry is to assure the maintenance of strong defense capabilities. However, if the defense industry is largely responsive to international market factors, as opposed to national defense needs, it is questionable whether the industry can adequately maintain national defense capabilities.

Third, there is a fear that the wholesale export of state-of-the-art military technology may provide U.S. foes and potential foes with a military advantage, since if they can get their hands on this technology, they can learn much from it through reverse engineering. This point was driven home forcefully with the sudden military collapse of South Vietnam, at which time Communist forces got hold of large quantities of advanced American military technology. Similarly, the overthrow of the Shah placed some of America's most advanced military hardware into the hands of people who were anything but friendly to the United States.

Finally, there is concern that the proliferation of American arms throughout the world promotes global instability and misery. The argument made here is that the acquisition of weapons will increase the level of destructive capacity in many parts of the world and may even encourage military adventurism. Some offer the rebuttal that arms sales can have a stabilizing effect, particularly when they prevent a country in a region from obtaining a preponderant military advantage over its neighbors.

The principal conclusion one can draw from a review of U.S. military sales policy since World War II is that it has fluctuated over the years in response both to external events (for example, the rise and decline of bipolarity) and internal factors (for example, Carter's concern about the proliferation of arms).

As arms systems become more costly and complex, there will be a compelling rationale for the United States to engage increasingly in coproduction agreements with its allies. The trend toward coproduction will be discussed in detail later in this chapter.

West Germany

In the first half of the twentieth century, Germany was recognized as one of the world's premier weapons manufacturers. The devastating effectiveness of these weapons was experienced strongly by the Allied powers in both World War I and II. After World War II, Germany was disarmed. The eastern portion of the country was occupied by the Soviet Union and the western portion was jointly administered by the Americans, British, and French. With the progression of the Cold War, West Germany was once again allowed to develop a defense force. The Bundeswehr (German army) was established in 1955 in order to provide a German contribution to NATO forces. Initially, the hardware employed by the German forces came from American, British, and French stocks. After a while, weapons were assembled and produced locally under Allied license. These licensing agreements eventually were converted into joint ventures and coproduction agreements made with British and French firms. Ultimately, German manufacturers came to assume a principal role in these weapons consortia.

With the bitter experiences of both world wars behind them, the West Germans have lost much of their taste for war-making. Article 26 of their constitution prohibits them from undertaking acts "tending to disturb peaceful relations between nations" and "preparation for aggressive war," as well as the "manufacturing, transport or marketing of weapons," except by government permission. In addition, the West Germans have one of the world's most stringent laws governing international arms sales, the Export Control Law of 1961 (amended in 1971 and 1978). This law has until recently effectively limited the sale of German arms to NATO countries and obvious friends, such as Australia, New Zealand, Japan, Switzerland, and Sweden. In order to help assure that West German manufactured arms do not wind up in the hands of other countries, a final-user clause is included in all of its arms sales agreements, stating that the arms will not be transferred to third parties.

Despite West German desires and attempts to control arms exports, West German arms, which are highly desirable for their reliability and high quality, nonetheless occasionally wind up in the hands of third parties. This is in part due to the West Germans' inability to enforce the final-user clause in their arms agreements. For example, arms sales to Italy are unrestricted, and Italians can reexport German arms anywhere, in spite of the final-user clause. West German arms can also be exported to third parties by West Germany's partners in joint ventures. Thus the French, coproducers with Germany of the Hot and Milan antitank missiles, can sell these missiles to third parties.

In recent times, pressures have grown to increase arms exports in West Germany. Clearly, the lucrative international arms trade would assure handsome profits for German arms manufacturers. But pressures also come from other sources as well. For example, a Saudi inquiry about the possibility of purchasing three hundred Leopard II tanks could not be dismissed out of hand, since Saudi Arabia is a major German creditor and a principal supplier of oil.[6]

In absolute terms, West German arms exports are not small. As table 11-3 showed, West Germany is the fifth largest arms exporter in the world. But only 0.5 percent of West German exports are arms, a far smaller fraction than the United States' 3.1 percent, Britain's 1.1 percent, or France's 1.4 percent. The West Germans clearly have the talent and capability to sell considerably more arms in the world marketplace.

Small Countries

Small countries that desire to produce arms for national defense face particularly strong pressure to export weapons, because the domestic demand for arms output is often insufficient to sustain a viable defense industry. However, the sophistication of the arms they produce, coupled with other factors, such as availability of spare parts, maintenance schedules, and distribution channels, will play a strong role in determining the nature and success of their export efforts.

Sweden is a country with only 4 percent of the U.S. population and 4 percent of the U.S. GNP, yet it is basically self-sufficient in arms production. Furthermore, Swedish arms, such as the Viggen jet aircraft, are highly regarded internationally. Swedish exports of arms could help underwrite the cost of its defense industry. However, the Swedes, whose views on political neutrality and arms control are well known, limit their arms exports to a handful of countries, such as Switzerland, and then only when they have assurances that the arms will be used for defensive purposes. Only 0.4 percent of Swedish exports in a year are arms.

Israel is another small country that has a substantial arms industry. Israel spends a larger portion of its GNP on defense than any other country. Throughout the 1970s, some 18 to 34 percent of the Israeli GNP was dedicated to military expenditures! Israel's ability to finance its defense effort is made possible by military assistance from the United States. However, this is chiefly tied-aid, requiring Israel to purchase weapons from the United States rather than helping the country to sustain domestic arms production. Consequently, Israel actively seeks foreign arms sales to keep its domestic arms industry viable. In recent years, a hefty 2 to 6 percent of Israel's total exports have been military hardware.

South Korea has a substantial defense industry, comprised of eighty companies in 1981, producing a wide range of military hardware under license from American sources. Korea moved to increase its domestic arms production in the mid-1970s, after the Nixon Doctrine was announced and after the Carter administration publicly entertained the possibility of a hasty withdrawal of American troops from the Korean peninsula. As with the case of other small countries committed to maintaining a domestic defense industry, Korea has sought some relief from the defense burden through exports. It has been reasonably successful in making military exports to its fellow Asian countries who are members of the Association of Southeast Asian Nations (ASEAN) and who are bolstering their defenses against what they perceive as a threatening posture by the Vietnamese.[7] In 1979 South Korea exported a substantial $180 million in arms.

Technology Transfer

When a country purchases antitank missiles from an arms supplier, or has its personnel trained abroad in flight school, or enters into a joint venture to produce supersonic fighters, or manufactures small arms under license, technology is being transferred. In most respects, the mechanisms for the transfer of military technology are the same as those associated with the transfer of civilian technology. However, in one important respect there is a difference, particularly in regard to the most advanced military technology. That is, while technology in the civilian sector is generally transferred in accordance with market forces and the decisions of individual firms that set the conditions for the transfer, technology transfer in the defense area is often regulated by governments. In deciding whether a given arms technology should be exported, a government weighs the transfer according to a number of criteria: the impact of the transfer on national security, its effect on the economy, its implications for foreign policy, and so on. If for whatever reason the government determines that the arms transfer is detrimental to the country, the government can forbid it.

The most common mechanism for obtaining military technology is purchasing it off the shelf. That is, arms purchasers generally buy hardware just the way it comes off the production line. Modifications may be made to the hardware, as when a fighter is fitted with larger than standard fuel tanks, but custom work is the exception rather than the rule. It should be recalled from the earlier discussion of technology-transfer mechanisms that off-the-shelf goods have technology embedded in them and consequently the recipient may learn very little from them.

The straightforward licensing of arms technology is also common, particularly for low and intermediate level technology. South Korea's license from Colt to produce M-16 rifles is an example of this. Governments are obviously cautious about licensing out advanced technologies, since it may involve the loss of important arms-technology secrets to a foe. In general, advanced military technologies, are licensed only to close, highly trusted allies if they are licensed at all.

In recent years there has been a trend toward increased coproduction of advanced military hardware, paralleling a similar trend in the civilian sector. As suggested in an earlier chapter, increases in civilian joint ventures and coproduction agreements are largely a consequence of changes in the international business environment, including: an increased willingness of firms to share technology, the high cost of financing new ventures, increases in international competitiveness and related improvements in the bargaining positions of technology recipients, foreign investment laws requiring national equity-shares of new ventures, and the feeling that joint ventures can be a useful tool for penetrating inaccessible markets.

Of the items listed here, the most significant in the military area are the desire for cost sharing and increased competitiveness in arms markets. The desire to share the costs of developing and deploying new weapons systems is understandable, particularly in this age of astronomical costs for complex systems. Cost sharing also entails skills sharing, so that coproduction agreements can take advantage of the strengths of the partners. This cost-sharing motivation has been particularly strong in Western Europe, and has led NATO members to undertake such joint ventures as production of the Jaguar fighter (United Kingdom and France), the Tornado jet (United Kingdom, Italy, and Germany), the Roland missile system (France and Germany), and the Hot and Milan antitank missiles (France and Germany).

With increased competition in international arms markets, weapons recipients that want to develop a domestic arms industry are able to ask for, and occasionally obtain, coproduction agreements, since arms suppliers may fear that if they do not strike such a deal, the recipient will go to another supplier. Thus, Brazil has been able to develop a substantial arms industry through coproduction agreements with Germany, France, Italy, and Belgium.

Another force that has contributed to the increase in coproduction agreements has been the requirement that the NATO allies standardize their weapons systems. At one time, standardization was maintained simply through the purchase of a common body of U.S. manufactured weapons. The reliance on U.S. weapons naturally began after world War II, when the United States made available to its allies at low cost weapons that would enable them to shore up their defenses against the Soviet menace. The ample supply of cut-rate weapons naturally caused NATO to think American in its arms purchases. However, replacements were expensive and led to a reconsideration by NATO members of their dependence on U.S. arms. They began developing their own substantial weapons industries and the resulting products led to a serious lack of standardization and interchangability of weapons in the NATO forces. It was clear that if NATO was to match the Soviet-bloc forces effectively, its own forces would need standardized weapons. However, the NATO countries were unwilling to return to a dependence on U.S. arms. They demanded what was termed two-way-street cooperative procurements within NATO, where NATO forces would use standard weapons of both American and European origin. Their position was buttressed in the United States by the passage of the Culver-Nunn amendment to the Department of Defense Appropriation Act of 1977. This amendment called for a degree of standardization of U.S. equipment used in Europe and recognized that this could be carried out through two-way-street procurements. An obvious way this could be effected was through coproduction agreements that assured Americans that, as the principal opponent of the USSR, they would have substantial control over the development of weapons systems that would dovetail with their defense needs; and that also enabled European countries to get a piece of the action.

Clearly, with licensing agreements, joint ventures, and coproduction agreements, more technology is transferred than with off-the-shelf purchases. These agreements entail the transfer of engineering specifictions, training of personnel, employment of high-technology equipment, and so on, while with off-the-shelf purchases the recipient merely acquires the end product of someone else's technology capabilities.

The overall benefits of arms technology transfer to the recipients are questionable. This is a consequence of the fact that military technologies are generally of limited economic value. Arms are final consumption items, not intermediate goods that generate more material production. For poorer countries especially, the purchase of military technologies may help deplete scarce hard currency reserves, reserves that could have purchased civilian technologies that would have led to increased skills acquisition and production capabilities. Certainly, when the recipient coproduces an item under an agreement with the technology donor, it will often acquire skills that can be applied in the civilian area as well. But this is an inefficient way to acquire needed skills. A country that wishes to obtain airplane manufacturing skills would do better to enter an agreement with a technology donor to produce civilian aircraft, for which there is a broad domestic and international market, than jet fighters, whose sales could never compete satisfactorily with sales of U.S., French, Swedish, or Soviet jets.

U.S. Export Controls

It is recognized by countries that produce and sell arms that arms sales can have effects that run contrary to the national interest. In particular, four possible negative consequences of arms sales stand out:

1. they may enable an opponent to obtain technical information detrimental to national security.
2. they may help create arms industries that ultimately will compete against domestic industries for arms sales to third parties.
3. they may lead to increased dependence on foreign military sales.
4. they may contribute to instability in a region where the introduction of new arms upsets the local balance of power.

Of special concern to the United States and its allies is the first of these effects: that arms sales by these countries may somehow enable the Soviet Union to increase its military capabilities. This concern has led the Western countries and Japan to attempt to restrict their overseas arms sales so as to keep critical technologies out of the hands of the Eastern block countries.

The United States has instituted some form of export controls over military sales continuously since the outbreak of World War II, when the Trading with the Enemy Act proscribed exports to the Axis powers. The outbreak of the Cold War, commencing immediately upon the termination of World War II, saw mandatory licensing of all exports to the USSR. The Export Control Act of 1949 formalized controls over goods sold to the USSR. The purpose of export controls at this time was to keep products that would strengthen Soviet economic and military capabilities out of their hands.

In order to coordinate the U.S. approach with that of its Allies in Europe, a Consultative Group on Export Controls was established in 1949. The operating arm of this group was the Coordinating Committee (COCOM), which ultimately was comprised of the NATO countries (less Iceland) and Japan. COCOM maintained a list of goods which, because of their military value, could not be exported to the Communist countries. The Consultative Group eventually dissolved, but COCOM still functions to this day.

Attempts to coordinate the export policies of the COCOM countries have never been fully satisfactory. The basic problem revolves around the stringency of the export controls. The Europeans favor looser controls than the Americans. They have a long tradition of trade with the Eastern European countries, so that such trade strikes them as quite natural. In addition, their outlook on the East is less colored by ideological considerations than the outlook of the Americans.

In order to back their position with a little muscle, the Americans passed the Mutual Defense Assistance Control Act of 1951 (Battle Act). This act attempted to impose some consistency on the export control policies of the United States and its allies. It empowered the U.S. president to terminate all military and economic assistance to countries that exported certain strategic goods to the Soviet Union and its satellites. Such sanctions have never been imposed.

By the end of the 1960s, much of the sharp edge of the Cold War had been dulled and the United States and Soviet Union found themselves moving in the direction of detente. It was in this relatively relaxed atmosphere that Congress passed the Export Control Act of 1969, which liberalized controls on the export of goods and technology to the Soviet Union. Bans on technologies that would help Soviet economic capabilities were lifted. However, prohibitions continued on exports of goods and technology that would strengthen Soviet military capabilities. The underlying premise of the revised American approach to trade with the USSR was that as the Soviets prospered economically, they would develop a stake in the existing world order and would work to preserve it rather than knock it down. Substantial quantities of U.S. technology began to flow to the Soviet Union in the 1970s.

It was not long before the Defense Department sounded the alarm that American technology, while purportedly of a civilian character, could easily be diverted for military purposes. For many technologies—for example, telecommunications equipment, integrated circuits, avionics, and aircraft engines—the distinction between civilian and military applications is a fine one. Indeed, time was to show that the Soviet Union had no compunctions in employing Western-provided civilian technology in its military operations. Many of the trucks used to transport Soviet troops and material into Afghanistan in the December 1979 Soviet invasion were built at the Kama River truck plant, which was constructed in the early 1970s with United States and other Western technology.

Amendments were made to the liberal 1969 Export Control Act in 1972 and 1974 in order to address some of the Defense Department's concerns, particularly those regarding the issue of dual-use technologies. In 1979 a new Export Administration Act was passed. The goals of the act are stated as follow:

It is the policy of the United States to use export controls only after full consideration of the impact upon the economy of the United States and only to the extent necessary

(A) to restrict the export of goods and technology which would make a significant contribution to the military potential of any other country or combination of countries which would prove detrimental to the national security of the United States;

(B) to restrict the export of goods and technology where necessary to further significantly the foreign policy of the United States or to fulfill its declared international obligations; and

(C) to restrict the export of goods where necessary to protect the domestic economy from the excessive drain of scarce materials and to reduce the serious inflationary impact of foreign demand.[8]

Thus, export controls can be imposed on national security, foreign policy, and economic grounds. Although the act established an Export Council consisting of a wide range of government agencies to provide advice on export policy, only three agencies play a dominant role in export control policy: the Department of Defense is the guiding force in military export policy, while the Department of State provides direction on export controls undertaken for foreign-policy reasons. The Department of Commerce is charged with the overall administration of the export control act and with providing guidance on the impact of exports on the U.S. economy.

Foreign countries are divided into several categories for export control purposes. Group Y is comprised of Communist countries with which the United States has normal, but cool, relations, including the Soviet Union. Group Z contains Communist countries with which the United States has poor relations, including North Korea, Vietnam, Cambodia, and Cuba. Export controls are most stringent with these countries. Group Q is Romania and Grow W contains Hungary and Poland. Group T consists of all the countries of the Western Hemisphere, except Cuba and Canada. Group V includes all other countries, except Canada. Canada fits into no specific category and controls on exports to Canada are vey lax.

With the exception of most goods shipped from the United States to Canada, all U.S. exports require licenses. There are two basic categories of licenses: general licenses and validated licenses. General licenses allow unrestricted export of goods of a nonstrategic nature that fall into one of more than a dozen categories identified by the Commerce Department. Some 90-95 percent of U.S. exports fall into this category. Validated licenses are granted for goods that might be proscribed according to their nature or destination. For example, goods in the commodity control list destined for shipment to the Soviet Union must be granted a validated license, otherwise they cannot be legally exported. Applications for validated licenses are examined carefully to determine whether their export is in compliance with the export control act. Some 5-10 percent of export licenses fall into this category.

U.S. export policy continues in a state of disarray, with Defense Department personnel taking a hardline view on the export of dual purpose goods and technologies, while Commerce Department personnel hold a more liberal trade outlook. Complicating matters is the general difficulty in coordinating U.S. policy with that of its COCOM partners. Initially, the United States was angered by what it perceived to be a cavalier attitude on the part of the other COCOM countries in dealing with export controls. But in recent years, the United States itself has been accused of this very shortcoming. This is seen in the share of exceptions to COCOM commodity restrictions granted to the United States. In 1962, the United States made only 2 out of a total 124 requests for exceptions brought before COCOM. By 1978,

62.5 percent of 1,050 requests for exceptions were of U.S. origin.[9] Currently, a large amount of attention is being devoted to creating a fair commodity control list that balances American trade needs against national security requirements.

Arms and the Developing World

It has already been mentioned that LDCs are major arms importers, obtaining their armaments for the most part from the developed countries in both the East and West. The share of global arms imports attributed to the LDCs has been growing steadily. In 1969, 66 percent of the world's total arms imports were made by LDCs, while in 1979 this figure had grown to 81 percent.

With the passage of the years the sophistication of the military technology imported by the LDCs has increased enormously. In the 1950s, most imported weaponry was of U.S. origin and World War II vintage. But growing defense industries in France, Italy, the United Kingdom, and East Europe made an increasingly broad array of hardware available to LDCs and introduced a strong element of competition into the international arms business. In the global scramble to sell arms, the LDCs found themselves largely in a buyer's market. In order to get the best terms in arms purchases, they would negotiate with several sellers simultaneously. Particularly important factors in making a sale were the kill-capacity of the weapons and the existence of good financial arrangements for arms payments.

The perils of not playing ball with Third World arms purchasers is amply illustrated in the following well-known case. Between 1974 and 1976, Peru made requests to purchase F-5 and A-4 aircraft from the United States. The U.S. government never acted upon these requests, so the Peruvians turned to the Russians and purchased 36 Sukloi Su-22 fighters from them for $250 million. This was the first purchase by a Latin American country (outside Cuba) of Soviet military aircraft.

Peru's increased airpower alarmed neighboring Ecuador, which consequently requested A-7 aircraft from the United States. The request was denied. Ecuadorian attempts to purchase Israeli Kfir C-2 fighter-bombers were similarly stymied because the Kfir contained a U.S. engine and the U.S. government would not allow the Israeli sale. Finally, towards the end of 1977, Ecuador purchased 58 Mirage F-1 fighters from France that were valued at $841 million.

Until recently, the United States has been the chief arms supplier to the Third World. Beginning in the late 1970s the estimated value of Soviet arms deliveries began to consistently exceed that of the United States. Table 11-4 shows the value of arms deliveries for the major exporters in 1974 and 1979 (measured in current dollars). The table shows that in the 1970s the USSR

Table 11-4
Arms Deliveries to the Third World, 1974 and 1979
(millions of dollars)

	1974	1979
Non-Communist	5,601	9,121
United States	4,021	5,711
France	480	755
United Kingdom	460	700
West Germany	180	230
Italy	130	425
Other	330	1,300
Communnist	2,880	8,860
Soviet Union	2,500	8,000

Source: Committee on Foreign Relations, U.S. Senate, "U.S. Conventional Arms Transfer Policy," (Washington, D.C.: U.S. Government Printing Office, June 1980).

dramatically increased arms deliveries, while in this same period U.S. deliveries increased more modestly.

The nature of the delivered arms is given in table 11-5. This table makes clear that LDCs are engaged in rather hard-hitting arms purchases and that their imports are not simply small arms. collectively, the arsenal being amassed in the Third World is very impressive indeed.

Because of the enormous expense involved in creating an indigenous arms industry, few LDCs have made substantial investments in domestic arms production. However, LDC arms production has shown signs of

Table 11-5
Weapons Delivered to the Third World, 1978-1980
(units of weapons sold)

Weapons Type	USSR	USA	Major West Europe
Tanks & self-propelled guns	6,022	3,036	795
Artillery	10,523	2,270	2,068
APCs & armored cars	7,642	7,167	2,615
Major surface combatants	25	22	34
Minor surface combatants	107	24	367
Submarines	5	3	12
Guided missile boats	43	0	17
Supersonic combat aircraft	1,761	510	251
Helicopters	676	187	916
Other aircraft	247	273	407
Surface-to-air missiles (SAMs)	13,654	5,620	2,049

Source: William H. Lewis, "Arms Transfers and the Third World," in U.S. Arms Control and Disarmament Agency, *World Military Expenditures and Arms Transfers, 1970-1979* (Washington, D.C., 1982), p. 29.

significant growth in recent years, as reflected in statistics on the LDC share of global arms exports. Excluding data for the People's Republic of China, we find that in 1969 0.5 percent of the dollar value of global arms exports could be attributed to LDCs, while 5.1 percent could be attributed to LDCs by 1979. (These figures, with China included, are 4.0 precent and 5.6 percent respectively.) The major developing countries engaged in arms production, as of 1980, were Brazil, India, Israel, South Africa, North Korea, and South Korea. These countries produce or assemble (through licensing arrangements) some rather sophisticated equipment, including aircraft, missiles, naval ships, and armored vehicles. Smaller, but significant, defense industries are supported in Argentina, Egypt, Indonesia, Pakistan, Singapore, Turkey, and Taiwan.[10]

Extensive Third World involvement in the purchase and production of weapons has substantial political, social, and economic implications. From the political standpoint, the weapons are often used for internal security purposes—that is, they are trained on domestic groups that pose a threat to the government. When weapons are purchased for defensive purposes, they can introduce a destabilizing element into the regional power configuration and lead to a localized arms race or even war. More often than not, the search for increased security through arms purchases is illusory. As the world's two superpowers painfully recognize, more weapons do not mean more security. Yet, a government that does not provide for the defense of the nation is failing in one of its principal duties. Even purportedly pacifist countries such as Sweden and Switzerland have substantial military capabilities. It is unfortunate that in regard to the purchase and production of weapons, the old adage appears to hold: Damned if you do, damned if you don't.

Socially, arms build-ups and the threat of war lead to a great deal of stress. Furthermore, weapons purchases and production divert resources away from vital social services. For example, in 1978 LDCs expended some $110 billion on all military expenditures, including $16.7 billion for arms imports. In contrast, they spent only $22.5 billion on public health and $67.0 billion on education. The emphasis on defense vis-a-vis social services is proportionately greater for LDCs than for the developed countries, which in 1978 spent $370.3 billion, $231.0 billion, and $374.2 billion on gross military expenditures, public health, and education respectively.

Perhaps the impact of arms transfers is most pronounced when viewed from the economic perspective. There are two major implications, under which many minor ones fall. First, there is the guns versus butter issue. All countries must wrestle with this matter, be they rich, middle income, or poor. However, its implications for poor countries is most poignant, because they are poor and the expenditure of tens of millions of dollars on arms by a country with a per-capita income of $150 would seem to be the most egregious kind of conspicuous consumption. In the overwhelming majority of cases, the purchase (or production) of weapons makes no positive contribution to the economy. On the contrary, it detracts from economic development. Consider that imported arms are generally purchased with hard currency, money that is also needed to purchase technology and industrial machinery from developed countries.

A second major implication pertains to those developing countries that have made a heavy commitment to establishing a major domestic weapons industry. While the existence of such an industry may increase an LDC's feeling of security by reducing its dependence upon outside arms-suppliers, it is unlikely that the industry will ever turn a substantial profit, particularly when one takes into account the opportunity costs of the venture. LDCs simply are unable to develop the technology that would enable their military products to compete satisfactorily in international markets, particularly in the case of major items such as fighter aircraft, missiles, and tanks. Yet, international sales are needed to underwrite some of the costs of the industry.

Notes

1. K. Kreilkamp, "Hindsight and the Real World of Science Policy," *Science Studies* 1 (1971):43-66.

2. W.H. Lewis, "Arms Transfers and the Third World," in *Arms Control and Disarmament Agency World Military Expenditures and Arms Transfers* 1969-1978 (Washington, D.C., 1982), p. 29.

3. J. Gansler, *The Defense Industry,* (Cambridge, Mass.: MIT Press, 1980), p. 209.

4. "U.S. Arms Policy Gives Major Role to Overseas Sales," *Wall Street Journal,* 10 July 1981, p. 25.

5. "Foreign Arms Support Straps Military," *Aviation Week and Space Technology* 108, 26 June 1978:72.

6. "Weapons Industry: A Reputation for Quality," *International Herald Tribune,* Monthly Supplement, April 1981, p. 168.

7. "South Korea's Role as a Military Supplier," *Business Week,* 29 June 1981, p. 72.

8. Pub. L. No. 96-72, 93 Stat. 503, Section 2402, Paragraph (2), (September 29, 1979).

9. T. Agres, "Moon Treaty Meets Resistance," *Industrial Research and Development,* (May 1980), p. 62.

10. ACDA, *World Military Expenditures,* p. 21.

Index

About the Author

J. Davidson Frame is director of the Program on Science, Technology, and Innovation at the School of Government and Business Administration, The George Washington University. Before joining the university faculty, he was vice-president of Computer Horizons, Inc., where he directed many projects examining both domestic and international scientific and technological activities. He has written extensively about international science and technology, as well as about the management of R&D. He received the Ph.D. from the School of International Service at The American University in Washington, D.C.

SOCIAL SCIENCE LIBRARY

Manor Road Building
Manor Road
Oxford OX1 3UQ
Tel: (2)71093 (enquiries and renewals)
http://www.ssl.ox.ac.uk

This is a NORMAL LOAN item.

We will email you a reminder before this item is due.

Please see http://www.ssl.ox.ac.uk/lending.html
for details on

- loan policies; these are also displayed on the notice boards and in our library guide.

- how to check when your books are due back.

- how to renew your books, including information on the maximum number of renewals.
Items may be renewed if not reserved by another reader. Items must be renewed before the library closes on the due date.

- level of fines; fines are charged on overdue books.

Please note that this item may be recalled during Term.